P9-BHT-411

The Generalized Reciprocal Method of Seismic Refraction Interpretation

by

Derecke Palmer
Department of Mineral Resources and Development
Sydney, N.S.W., Australia

Edited by

Kenneth B.S. Burke
Department of Geology
University of New Brunswick
Fredericton, N.B., Canada

ISBN 0-931830-14-1

Library of Congress Catalog Number: 80-52549

Society of Exploration Geophysicists
P.O. Box 3098, Tulsa, Oklahoma 74101

Published 1980

Printed in the United States of America

Table of Contents

	Page
Foreword	v
Abstract	vii
Chapter 1 Introduction	1
The seismic refraction method	1
Interpretation methods and the GRM	1
Chapter 2 Traveltime expressions	3
The seismic model	3
Traveltime expressions	3
Chapter 3 Refractor velocity determination	7
The velocity analysis function	7
Plane layers	7
Chapter 4 The time-depth function	13
Definition	13
The conventional time-depth	13
The delay time	13
Hales's method	14
The depth conversion factor	14
Time-depths near shotpoints	15
Continuous change of velocity with depth	15
Chapter 5 Synthetic models	17
Irregular refractor surface	17
Irregular ground surface	22
Irregular ground and refractor surface	25
Continuous change of velocity with depth	29
Detail of refractor definition	29
Chapter 6 Selection of XY-values	31
Small-scale irregularities	31
Determination of an optimum XY-value	32
Summary	34
Chapter 7 Undetected layers	37
The blind zone	37
Second events	38
Velocity inversions	39
Detection of hidden layers and velocity inversions	39
Summary	39

Chapter 8 Average velocity 41
  Definition 41
  Errors for the three-layer case 43
  Comparison with hidden layer errors 44
  Corrections for surface layers 45
  Synthetic example 46
  Summary 47
Chapter 9 Reciprocal time 49
  Reciprocal times for distant shots 49
  Corrections for errors in reciprocal times 50
  The intercept time 52
Chapter 10 The time section 53
  Uniqueness of the time section 53
  Consistency between data and interpretation 54
  Field data requirements 54
Chapter 11 An interpretation routine 55
  Data processing 55
  Interpretation 56
  Depth section 57
  Wavelength considerations 57
Chapter 12 A field example 59
  Geologic setting 59
  Seismic velocity stratification 63
  Initial interpretation 63
  Determination of *XY*-values 67
  Revised interpretation 67
  Small refractor irregularities 74
  Accuracy of depth estimates 74
  Refractor velocity variations 74
  Summary 81
Chapter 13 The future 83
Acknowledgments 85
References 87
Appendix A Traveltime data 90
Appendix B Welcome reef data 93
Index 103

Foreword

One has only to check a recent textbook on exploration geophysics to appreciate the fact that many methods have been proposed for the interpretation of seismic refraction data. In this monograph, Derecke Palmer describes a new comprehensive method of interpretation, the generalized reciprocal method (GRM), for which many of the previously presented methods are special cases. It also has the advantage of combining many of the better features of the individual methods.

The GRM is based upon the calculation of time depth and velocity function values from forward and reverse times of arrival at different geophone separations *XY* along a refraction profile. The reciprocal time from both directions of shooting is also required in the calculation.

One value of *XY* will be associated with arrivals from approximately the same point on the refractor, this optimum value of *XY* being determined from characteristic features of plotted time depth and velocity function values. Thus, the GRM combines the migration aspects of the delay time method (Barry, 1967) with the simplicity of computation of the reciprocal method of Hawkins (1961).

The time section produced by plotting of the time depths of the optimum *XY* separation provides the basic data for depth determinations and forms a convenient separation point between the data processing and interpretation stages. The time section is converted to a depth section using the best information available. A comparison of *XY* values calculated from this depth section with the optimum *XY* values used in data processing allows the detection of possible hidden layers in the overburden. An average overburden velocity can also be calculated from the optimum *XY* value. This permits accurate depth calculations to be made even in the presence of hidden layers and continuous changes in velocity with depth.

Thus the GRM offers many advantages over previously published methods of seismic refraction interpretation. The best way to appreciate its capabilities is to try it out on one of your own seismic refraction data sets, perhaps the one that gave you most trouble in the past. I do not think you will be disappointed with this new approach to seismic refraction interpretation.

November 15, 1979

KENNETH B. S. BURKE
Special Editor

Abstract

The generalized reciprocal method (GRM) is a technique for delineating undulating refractors, at any depth, from in-line seismic refraction data consisting of forward and reverse traveltimes.

The arrival times at two geophone positions, separated by a variable distance XY, are used in refractor velocity analysis and time-depth calculations. At the optimum XY separation, the rays to each geophone emerge from near the same point on the refractor, and the refractor velocity analysis and time-depths are the most detailed.

Perpendicular thicknesses are obtained from time-depths and the depth conversion factor. Loci, rather than actual depth points, are determined, and the surface of the refractor is taken as the envelope of these loci. The depth conversion factor is independent of dip for angles up to about 20 degrees; thus depth calculations to an undulating refractor are particularly convenient, even when the overlying strata have velocity gradients.

The presence of undetected layers can be inferred when the observed optimum XY-value differs from that derived from the computed depth section.

The optimum XY-value can be used to form an average velocity which permits accurate depth calculations with commonly encountered velocity contrasts.

Introduction

The seismic refraction method

Conventional seismic refraction methods aim to determine the spatial distribution of seismic wave velocities in the subsurface. Seismic wave velocities can be related to such geologic and petrophysical parameters as rock type, porosity, weathering, jointing, water saturation, and elasticity. Seismic refraction methods have been applied to petroleum exploration, the search for groundwater, the investigation of engineering sites, the exploration of alluvial deposits, and the correction of weathering effects for seismic reflection surveys.

Descriptions of instrumentation and field operations, as well as the fundamental principles based on raypath theory, can be found in Dobrin (1976) and Telford et al (1976). An introduction to the theoretical development using the wave equation was presented in Grant and West (1965).

Interpretation methods and the GRM

There are numerous interpretation methods, ranging from the very simple in basic assumptions and ease of use to the complex. Dobrin (1976) gives an excellent summary of most of the published methods, and, together with the outstanding volume on seismic refraction prospecting edited by Musgrave (1967), provides an ideal starting point for those who seek familiarity with one of the most challenging fields of geophysics. The aim of this monograph is to propose a new interpretation method, the generalized reciprocal method (GRM), which has many advantages compared with the previously published methods.

The GRM can define layers with varying thicknesses and seismic velocities, unlike the conventional intercept time method (Ewing et al, 1939; Dooley, 1952; Adachi, 1954; Mota, 1954), or the critical distance method (Heiland, 1963, p. 527). However, a redefinition of the intercept time in chapter 9 does permit partial extension of that method to irregular layers.

Like the conventional reciprocal method (Hagiwara and Omote, 1939; Hawkins, 1961), the GRM uses both forward and reverse arrival times and is relatively insensitive to dip angles up to about 20 degrees (Palmer, 1974). As a result, depth calculations to an undulating refractor are particularly convenient, even when the overlying strata have velocity gradients.

However, the conventional reciprocal method smooths refractor irregularities because it assumes a plane refractor between the points of emergence of the forward and reverse rays. On the other hand, the delay time method (Gardner, 1939, 1967; Barry, 1967) does not exhibit the same smoothing properties, but it is sensitive to dip angles as small as 5 degrees (Palmer, 1974).

Like the delay time method, the GRM employs the principle of migration. Arrival times at two geophones, separated by what is termed the *XY*-distance, are used in refractor velocity analysis and time-depth calculations. A range of *XY* spacings is used, and the optimum value is selected using various tests associated with the method. At the optimum *XY* spacing, the forward and reverse rays emerge from near the same point on the refractor, so the refractor need only be plane over a very small interval.

The determination of the optimum *XY*-value is not essential for accurate depth calculations. However, an accurate value is critical for the detection and accommodation of hidden layers and velocity inversions, and for the derivation of an average velocity. This average velocity can be very useful; it is analogous in any ways to the RMS velocity of reflection methods.

The GRM is ideally suited to processing by digital computer. Although each refractor velocity analysis and time-depth calculation involves only a few simple arithmetic operations, there can be many such calculations. However, the GRM is not an iterative method, such as the methods described by Scott (1973) and Singh (1978), so digital computers are not essential.

The processing routine used with the GRM offers significant advantages in the management of time, costs, and expertise (Palmer, 1979). Processing can be carried out before and independently of detailed interpretation, unlike methods such as wavefront construction (Thornburgh, 1939; Rockwell, 1967). Therefore, considerable savings can be achieved by using copies of plots of the data and processing data for both the initial interpretation, and for any subsequent reevaluation, when more information becomes available.

Chapter 2
Traveltime expressions

The seismic model

The two-dimensional model chosen for the derivation of GRM parameters consists of multiple plane-dipping layers with constant seismic velocities. This model does not represent the limitation of the method but has been selected for mathematical convenience.

There are several methods of specifying depths and raypath parameters. For example, Dooley (1952) and Adachi (1954) used vertical depths. However, a more convenient approach was used by Ewing et al (1939) and Mota (1954), who specified thicknesses normal to the refractor surface. The surface of the refractor is taken as the envelope of arcs of appropriate radii; hence loci, rather than actual depth points, are determined. The dip information, which is not always readily determined for undulating refractors, is recovered with the construction of the envelope. Therefore, the common assumption of the seismic profile being normal to the strike of the refractor is not necessary. This also permits convenient extension to three-dimensional (3-D) analysis.

Another advantage of the specification of Ewing et al (1939) is the symmetry of the resulting mathematical expressions. This symmetry, in turn, results in a depth conversion factor which is insensitive to dip angles up to about 20 degrees. Therefore, depth calculations are extremely convenient, both when the refractor is undulating and when there are complex velocity distributions above the refractor. These advantages are not readily achieved with the expressions of Mota (1954), who also used perpendicular thicknesses.

Accordingly, layer thicknesses and angles of incidence used in the following analysis are similar to those used by Ewing et al (1939). However, dip angles will be specified differently, with absolute, rather than relative, values being used.

Traveltime expressions

Traveltime expressions for multiple plane-dipping layers have been derived by Adachi (1954), but they are not suitable for derivation of GRM parameters.

Although the derivation can be applied to any number of layers, a four-layer case will be considered for convenience (Figure 1). Consider a plane wave passing through A, with the wavefront normal to AD. The wavefront will be normal to the raypath $ADFST$ at all times. The velocity of the wavefront is

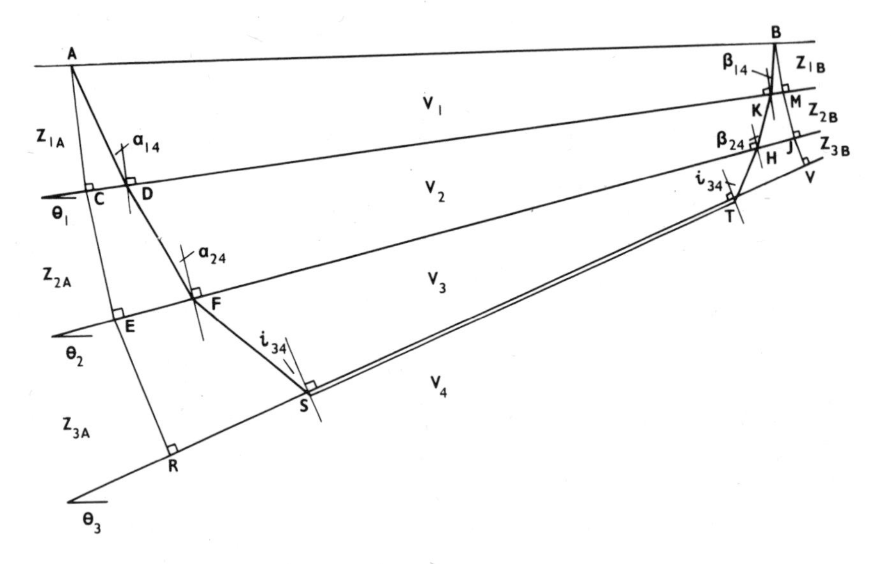

FIG. 1. Model and raypath parameters used in the derivation of the traveltime expression for the four-layer case.

the seismic velocity of each layer as measured along the raypath. This same plane wave will also pass through A, C, E, R, and S. At R and S it will be critically refracted.

When the wavefront arrives at C, it will have traveled a distance $AC \cos \alpha_{14}$ along AD, at a velocity of V_1. Hence, the traveltime from A to C is

$$Z_{1A} \cos \alpha_{14}/ V_1.$$

Similarly, the traveltime from C to E is

$$Z_{2A} \cos \alpha_{24}/ V_2,$$

and the traveltime from E to R is

$$Z_{3A} \cos i_{34}/ V_3.$$

At R, the wavefront is critically refracted, and the traveltime from R to S is

$$RS/ V_4.$$

Hence the traveltime from A to S is

$$\sum_{j=1}^{3} Z_{jA} \cos \alpha_{j4}/ V_j + RS/ V_4.$$

A similar result can be derived for the traveltime from T to B, i.e.,

$$\sum_{j=1}^{3} Z_{jB} \cos \beta_{j4}/ V_j + TV/ V_4.$$

Therefore, the traveltime from A to B is

$$t_{AB} = \sum_{j=1}^{3} (Z_{jA} \cos \alpha_{j4} + Z_{jB} \cos \beta_{j4}) / V_j + RV / V_4.$$

Furthermore, it can be readily shown that

$$RV = AB \cos \theta_1 \cos (\theta_2 - \theta_1) \cos (\theta_3 - \theta_2).$$

Therefore, the traveltime from A to B is

$$t_{AB} = \sum_{j=1}^{3} (Z_{jA} \cos \alpha_{j4} + Z_{jB} \cos \beta_{j4}) / V_j + AB \cos \theta_1 \cos (\theta_2 - \theta_1) \cos (\theta_3 - \theta_2) / V_4.$$

The expression for the multiple plane-dipping layer case (n layers) can therefore be shown to be

$$t_{AB} = \sum_{j=1}^{n-1} (Z_{jA} \cos \alpha_{jn} + Z_{jB} \cos \beta_{jn}) / V_j + AB \cos \theta_1 \cos (\theta_2 - \theta_1) \ldots \cos (\theta_{n-1} - \theta_{n-2}) / V_n. \quad (1)$$

For zero dip angles, equation (1) readily reduces to the horizontal plane layer equation of Jakosky [1950, p. 753, eq. (103)].

Chapter 3
Refractor velocity determination

The velocity analysis function

When the subsurface can be approximated with plane layers and uniform velocities, the refractor velocity can be obtained from the forward and reverse apparent velocities on the time-distance plot, together with the overlying velocities, using equation (1) of Ewing et al (1939, p. 265).

If these approximations cannot be made, or when the velocities of all layers above the refractor are not known, it is still possible to obtain a reasonable estimate of the refractor velocity by the following approach. Using the symbols of Figure 2, the velocity analysis function t_v is defined by the equation

$$t_v = (t_{AY} - t_{BX} + t_{AB})/2. \qquad (2)$$

The value of this function is referred to G, which is midway between X and Y.

In routine interpretation, values of t_v, calculated using equation (2), are plotted against distance for different XY-values. By a series of tests to be described in chapter 6, an optimum value of XY is selected, and a refractor velocity is taken as the inverse slope of a line fitted to the t_v values for the optimum XY.

For the special case of XY equal to zero, equation (2) reduces to equation (7) of Hawkins (1961, p. 809). It is similar to the minus term in the plus-minus method (Hagedoorn, 1959). The velocity analysis formula quoted by Scott (1973, p. 275) is a least-squares fit of data values which are mathematically similar to equation (2), but with XY equal to zero.

Plane layers

One method of testing the efficacy of determining refractor velocities with equation (2) is to apply the equation to a plane-layer case.

From Figure 2, it can be shown that

$$\begin{aligned} Z_{jX} &= Z_{jA} - (AG - GX)\, S(\theta_j), \\ Z_{jY} &= Z_{jA} - (AG + GY)\, S(\theta_j), \\ Z_{jP} &= Z_{jA} - GX\, S(\theta_j), \end{aligned} \qquad (3)$$

where

$S(\theta_j) = \cos\theta_1 \cos(\theta_2 - \theta_1) \ldots \cos(\theta_{j-1} - \theta_{j-2}) \sin(\theta_j - \theta_{j-1})$,

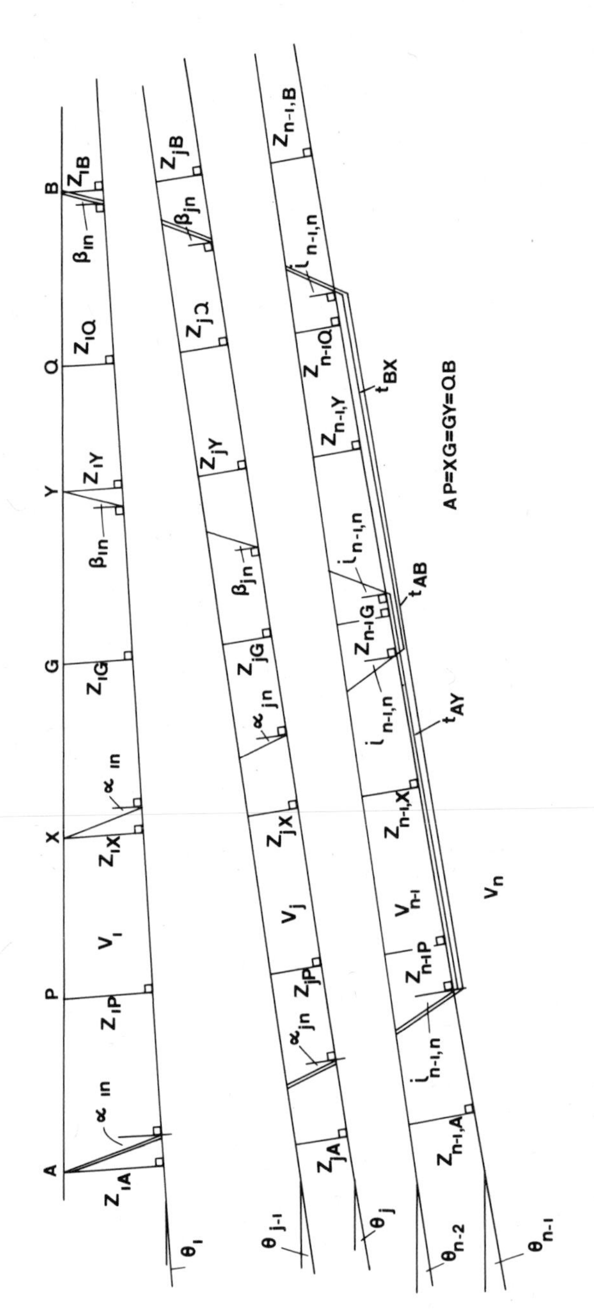

FIG. 2. Raypaths used in the definitions of the velocity analysis and time-depth functions.

and

$$S(\theta_1) = \sin \theta_1. \tag{4}$$

When the appropriate modifications of equation (1), together with equation (3), are substituted, equation (2) becomes

$$\begin{aligned} t_v = & \sum_{j=1}^{n-1} Z_{jP} (\cos \alpha_{jn} + \cos \beta_{jn})/2V_j \\ & + AG \, [\cos \theta_1 \cos (\theta_2 - \theta_1) \ldots \cos (\theta_{n-1} - \theta_{n-2})/V_n \\ & - \sum_{j=1}^{n-2} (\cos \beta_{jn} - \cos \alpha_{jn}) \, S(\theta_j)/2V_j] . \end{aligned} \tag{5}$$

Equation (5) represents a linear relation between t_v and the distance AG. The derivative of this function with respect to distance, i.e., the slope, is the coefficient of AG, and it will be defined as the inverse of an apparent velocity V'_n, i.e.,

$$\frac{d}{dx} t_v = 1/V'_n. \tag{6}$$

In routine interpretation, V'_n is usually assumed to be the correct refractor velocity. A measure of the accuracy of this assumption can be obtained by substituting the appropriate values taken from a fully defined model, such as that shown in Figure 3, into the coefficient of AG in equation (5). The angles of the raypaths to the normals of each interface are found by applying Snell's law.

The value for V'_4 is 4210 m/sec, which is about 5 percent greater than the correct refractor velocity. This error is quite acceptable, especially for such an extreme model with unusually large dip angles.

The refractor velocity estimate can be improved if a value of the refractor dip is obtained. If the difference in dip angles between successive layers is less than about 20 degrees, then the terms containing the expressions $S(\theta_j)$ generally have a negligible contribution to the coefficient of AG. Also for the same dip angle conditions

$$\begin{aligned} \cos (\theta_{n-1} - \theta_m) & = \cos [(\theta_{n-1} - \theta_{n-2}) + (\theta_{n-2} - \theta_m)] \\ & \simeq \cos (\theta_{n-1} - \theta_{n-2}) \cos [(\theta_{n-2} - \theta_{n-3}) + (\theta_{n-3} - \theta_m)] \\ & \approx \cos (\theta_{n-1} - \theta_{n-2}) \cos (\theta_{n-2} - \theta_{n-3}) \ldots \cos (\theta_{m+1} - \theta_m). \end{aligned} \tag{7}$$

Therefore, the coefficient of AG can be written

$$\frac{d}{dx} t_v \simeq \cos \theta_{n-1}/V_n. \tag{8}$$

Equations (6) and (8) can be combined to form

$$V_n \simeq V'_n \cos \theta_{n-1}. \tag{9}$$

Using equation (9), the improved estimate for the refractor velocity for the model in Figure 3 is 4066 m/sec (4210 cos 15). This value represents an error of 1.5 percent.

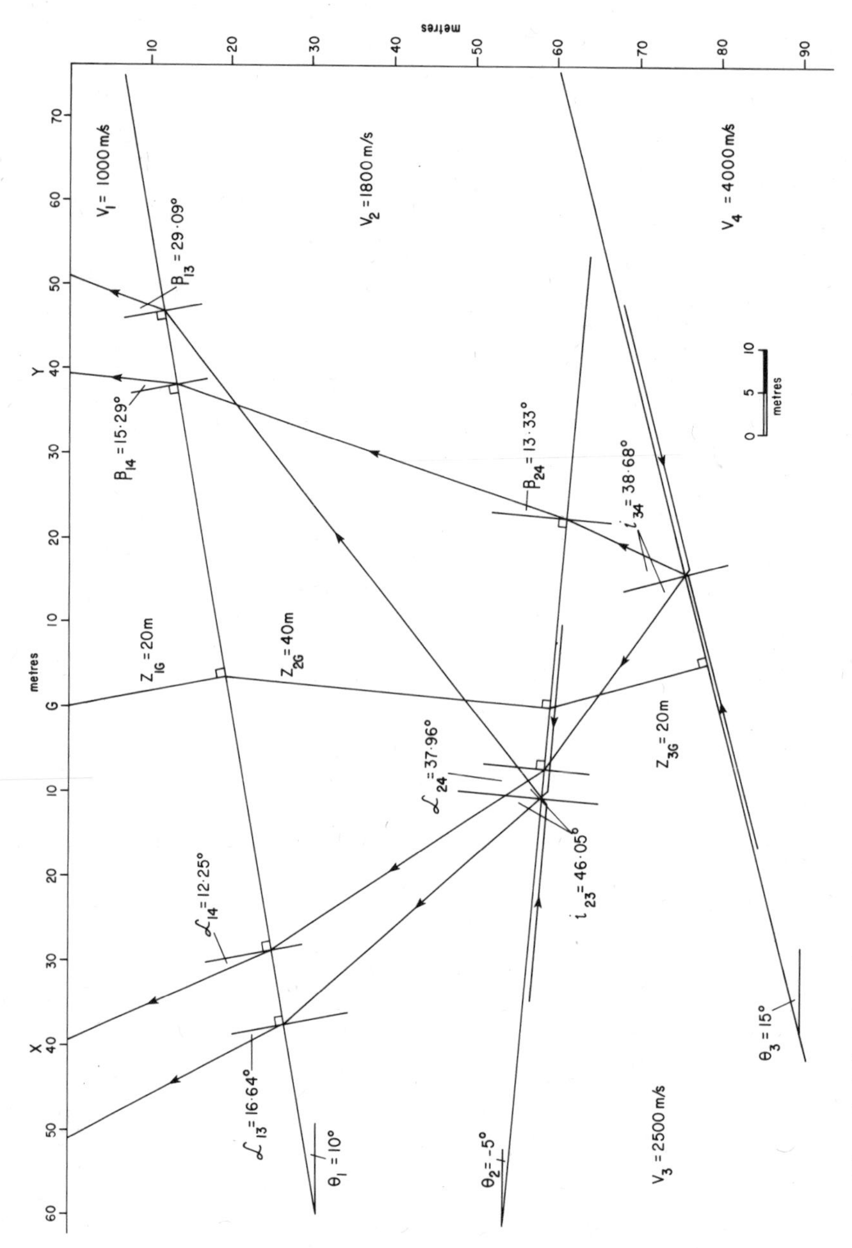

FIG. 3. Model and raypath parameters used to evaluate GRM functions for an extreme case.

Table 1. Errors introduced by the two velocity assumptions.

j	V_j' (m/sec)	Error (percent)	$V_j' \cos \theta_{j-1}$ (m/sec)	Error (percent)
1	1000	—	1000	—
2	1828	1½	1800	0
3	2578	3	2568	3
4	4210	5	4067	1½

In Table 1, all the apparent velocities, the velocities approximately corrected for dip, and the errors are presented. As Table 1 demonstrates, very good refractor velocity estimates can be made if a measure of the refractor dip angle is available. Such a value might be obtained from drill hole control or even from preliminary depth calculations. However, when corrections for refractor dip are not made, lateral variations in the refractor velocity of less than 5 per cent might not be significant.

Chapter 4
The time-depth function

Definition

Following the determination of the refractor velocity, the next step in defining undulating refractors is the formation of generalized time-depth functions at each geophone location. The generalized time-depth function in refraction interpretation corresponds with (but is not identical to) the one-way traveltime-depth function in reflection methods.

Using the symbols of Figure 2, the generalized time-depth t_G(hereafter referred to as "time-depth") at G is defined by the equation

$$t_G = [t_{AY} + t_{BX} - (t_{AB} + XY/V'_n)]/2. \tag{10}$$

The term V'_n is the apparent refractor velocity determined from the velocity analysis function.

Several special cases of the generalized time-depth can be derived, depending upon the XY separation used.

The conventional time-depth

For XY equal to zero, the conventional time-depth (Hagiwara and Omote, 1939, p. 127; Hawkins, 1961, p. 807, eq. 3; Dobrin, 1976, p. 218, eqs. 7–35, 7–36) is obtained. It is similar to the plus term in the plus-minus method (Hagedoorn, 1959; Hawkins, 1961, p. 814) and to a term in the method of differences (Heiland, 1963, p. 549, eq. 9–68).

For the calculation of the conventional time-depth, no knowledge of the refractor velocity is required.

The delay time

For XY selected such that the forward and reverse rays emerge from near the same point on the refractor, a result similar to the mean of the migrated forward and reverse delay times, as defined by Barry (1967, p. 348), is obtained.

The delay time method was first described by Gardner (1939, 1967), and has been developed by many others (Barthelmes, 1946; Wyrobek, 1956; Bernabini, 1965; Layat, 1967; Peraldi and Clement, 1972). Although all theoretical derivations assume negligible dip angles, the method is generally considered valid for dips less than 10 degrees. However, dips as small as 5 degrees result in large differences between the forward and reverse delay times (Palmer, 1974), and depth calculations can have significant errors. Furthermore, to calculate accurate offset distances [Sheriff, 1973, p. 154, Figure (130b)] for migration, both overburden velocities and dips are required, but are not always known. If the forward and reverse delay times do not migrate to the same position on the surface, then production of an average delay time curve will involve interpolation. Therefore, the benefits of precise migration are greatly reduced.

These errors can be overcome by applying a uniform migration and then taking a mean of the migrated delay times. This procedure is automatically achieved with the generalized time-depth when the XY-value is optimum, i.e., when the forward and reverse rays emerge from near the same point on the refractor. In addition, separation of geophone and shotpoint delay times, migration, and convergence corrections are effectively incorporated into a single operation. However, instead of locating the exact point of emergence on the refractor of the forward and reverse rays, a point midway between X and Y is used. The errors introduced by this approximation are negligible for dips less than 20 degrees.

Hales's method

Another method similar to the GRM is Hales's method (Hales, 1958; Woolley et al, 1967). The method essentially determines a critical reflection time. This is equivalent to a generalized time-depth when the forward and reverse rays emerge from near the same point on the refractor, but without the XY/V'_n term in equation (10).

However, unless the separation between the forward and reverse times can be determined accurately, this critical reflection time can be considerably in error. The determination of the separation by anomaly correlation (Woolley et al, 1967, p. 280) can be difficult (see chapter 6), while the use of average velocities (Hales, 1958, p. 288) is inaccurate when hidden layers occur.

The generalized time-depth overcomes the problems of requiring an accurate separation by inclusion of the term XY/V'_n. This term allows equation (12) below, which relates time-depths with depths, to be valid, irrespective of the XY separation.

The depth conversion factor

From Figure 2, it can be shown that

$$Z_{jX} = Z_{jG} + GXS(\theta_j),$$

and

$$Z_{jY} = Z_{jG} - GXS(\theta_j), \tag{11}$$

where $S(\theta_j)$ is defined by equation (4). Therefore, for plane layers, equation (10) reduces to

$$t_G = \sum_{j=1}^{n-1} Z_{jG}\,(\cos\alpha_{jn} + \cos\beta_{jn})/2V_j. \tag{12}$$

Table 2. Errors introduced by using equation 14.

j	$2V_j/(\cos\alpha_{j4} + \cos\beta_{j4})$ m/sec	$V'_4 V'_j/(V'^2_4 - V'^2_j)^{1/2}$ m/sec	Error (percent)
1	1030	1029	—
2	2044	2029	1
3	3202	3261	2

Equation (12), relating time-depths and thicknesses, contains the expression

$$V_{jn} = 2V_j/(\cos\alpha_{jn} + \cos\beta_{jn}). \qquad (13)$$

V_{jn} will be termed the depth conversion factor. For zero dips, it is equivalent to the A function of Meidav (1960, p. 1049–1051), the depth conversion factor of Hawkins (1961, p. 807, 808), twice the G factor of Stulken (1967, p. 312), and twice the variable W of Chan (1968).

In general, α_{jn} and β_{jn} depend upon dip angles, which can vary randomly, and so exact calculation of V_{jn} is usually not possible. Normally the effects of dip angles are ignored, and equation (13) is approximated with equation (14), viz,

$$V_{jn} \simeq V'_n V'_j / [V'^2_n - V'^2_j]^{1/2}. \qquad (14)$$

The velocities used in equation (14) are those measured with equation (6).

A measure of the error in this approximation can be obtained by evaluating equations (13) and (14) with variables taken from the model in Figure 3 and Table 1. From Table 2, it is clear that the approximation of equation (14) is valid.

Time-depths near shotpoints

From equation (5), it can be seen that the intercept of the velocity analysis function at the shotpoint is the time-depth [compare with equation (12)] at a point displaced $XY/2$ (i.e., GX) from that shotpoint, toward the reverse shotpoint, i.e.,

$$[t_v]_{x=0} = \sum_{j=1}^{n-1} Z_{jP}(\cos\alpha_{jn} + \cos\beta_{jn})/2V_j. \qquad (15)$$

Hence when $XY/2$ is approximately equal to the offset distance, equation (15) provides a time-depth approximately where the ray from the shotpoint is critically refracted.

In general, it is advisable to compute velocity analysis functions for both the forward and reverse traveltimes, so that time-depths near each shotpoint can be obtained.

Continuous change of velocity with depth

Rigorous treatments to determine the depth to a dipping refractor, overlain by a layer with velocity varying continually with depth, are quite complex (Laski,

1973, 1978). As a result, most treatments have assumed horizontal refractors (Goguel, 1951; Kaufman, 1953; Duska, 1963; Hollister, 1967; Musgrave and Bratton, 1967) or have used graphical methods (Dix, 1952, p. 257; Olhovich, 1959; Northwood, 1967). However, the GRM is particularly suited to dealing with complex velocity distributions because the depth conversion factor is relatively insensitive to dip angles.

For the velocity to vary as a function of depth only,

$$\theta_j = 0,$$

and

$$V_j = V(z). \tag{16}$$

With the substitution of equations (14) and (16), and converting the summation into an integration, equation (12) becomes

$$t_G = \int_0^z \frac{\sqrt{V_n'^2 - V^2(z)}}{V_n' \, V(z)} \, dz. \tag{17}$$

Therefore, with the substitution of the true refractor velocity by that determined by the velocity analysis function, the relationship between the time-depth and the thickness is the same for both dipping and horizontal refractors.

Of all the functions used to describe continuous changes of velocity with depth, perhaps the most convenient is the Evjen equation (Evjen, 1967), viz.,

$$V(z) = V_1 (1 + az)^{1/q}, \tag{18}$$

where

$$q \geqslant 1 \quad \text{and} \quad a > 0.$$

The Evjen equation is one of very few equations which satisfy the important requirements of capability of being integrated and easy construction of trajectories. When q is 1, the familiar linear velocity function is obtained, and when q is 2, the parabolic function results.

The theoretical treatments of Iida (1939), Gassmann (1951, 1953), Brandt, (1955), Paterson (1956), and Berry (1959), as well as the experimental studies of Faust (1951, 1953) and Acheson (1963), indicate that q equal to 6 is probably the most appropriate value for porous media.

When q is greater than 2, the rate of increase of velocity decreases rapidly with depth, so velocities are well-behaved physically.

Substitution of equation (18) into equation (17) yields

$$t_G = \frac{q V_n'^{q-1}}{a V_1^q} \int_{\theta_1}^{\theta_2} (\sin^{q-2}\theta - \sin^q \theta) \, d\theta, \tag{19}$$

where

$$\theta_1 = \sin^{-1}(V_1 / V_n'),$$

and

$$\theta_2 = \sin^{-1}[V_1 (1 + az)^{1/q} / V_n'].$$

An empirical relationship between t_G and z can be obtained by replacement of the integral with a summation approximation, after values of q, a, V_n', and V_1 have been determined by other methods.

Chapter 5
Synthetic models

In the previous two chapters, the GRM parameters of velocity analysis function and generalized time-depth were defined, and then successfully applied to an extreme model with steeply dipping interfaces. Although plane-layer conditions are not uncommon, irregular layers are more usual and are generally of more interest. The following synthetic models permit the GRM to be examined in a variety of cases where departures from plane uniform layering occur.

In the time-depth graphs to follow (Figures 7, 22, 38, 44, and 50), the upper sets of points (circles) are the values for zero XY, the crosses are the values for a 5-m XY-value, and so on. Also, to avoid overplotting of points for various XY-values, each set of calculations uses a different reciprocal time. This results in a simple vertical displacement of plotted values which can be readily corrected if required (see chapter 9).

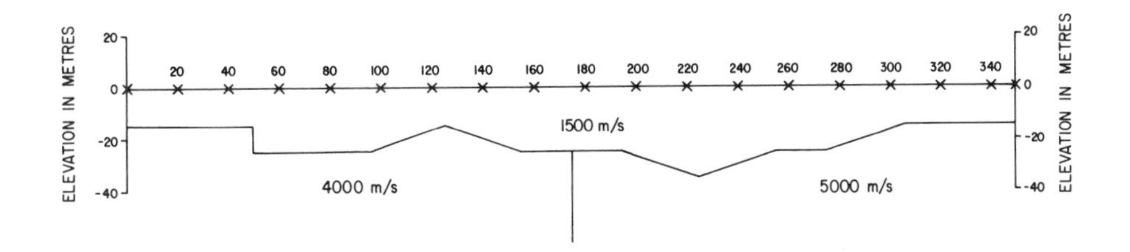

FIG. 4. Model with a plane horizontal ground surface and a highly irregular refractor. The vertical and horizontal scales are equal.

Irregular refractor surface

Perhaps the model of most interest is the irregular refractor. The model shown in Figure 4 has a plane horizontal ground surface and a highly irregular refractor with dips on the sloping surfaces of approximately 18 degrees. The first arrival times for this model, shown in Figure 5, were obtained by wavefront construction, since this method conveniently accommodates dipping refractors, interfering head waves, and diffraction (Thornburgh, 1939; Rockwell, 1967; Palmer, 1974). The traveltimes are also listed in Appendix A.

In Figure 6, the velocity analysis function is plotted for XY-values ranging from 0 to 30 m. For a 20-m XY spacing, the velocity analysis data fall very close to two straight lines, which provide velocities within 1 percent of the correct values.

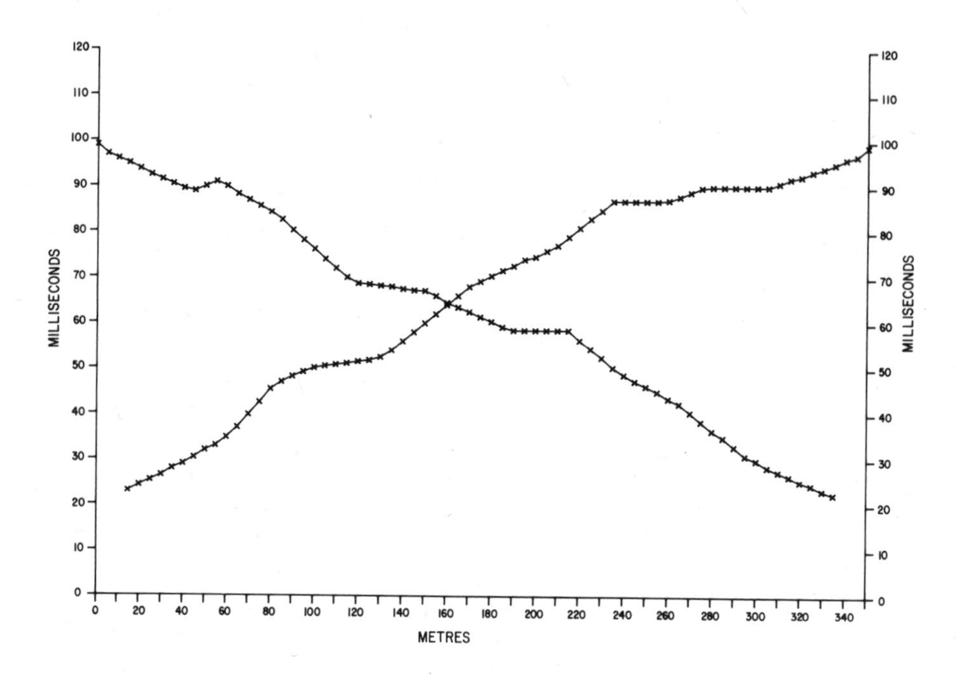

FIG. 5. Traveltime curves derived from the model in Figure 4.

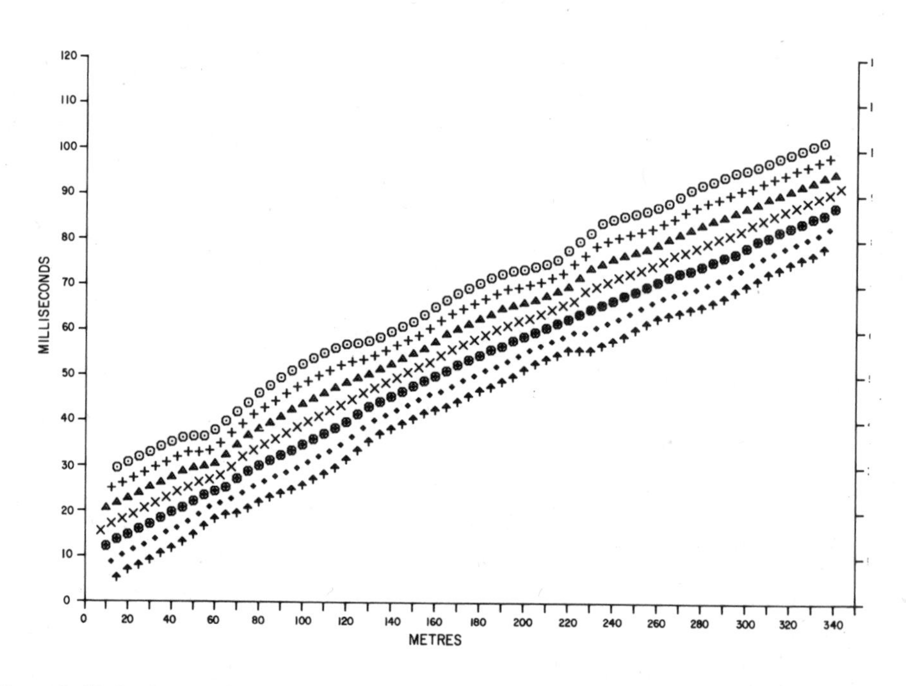

FIG. 6. Velocity analysis functions for XY-values from 0 to 30 m, derived from the traveltime data in Figure 5.

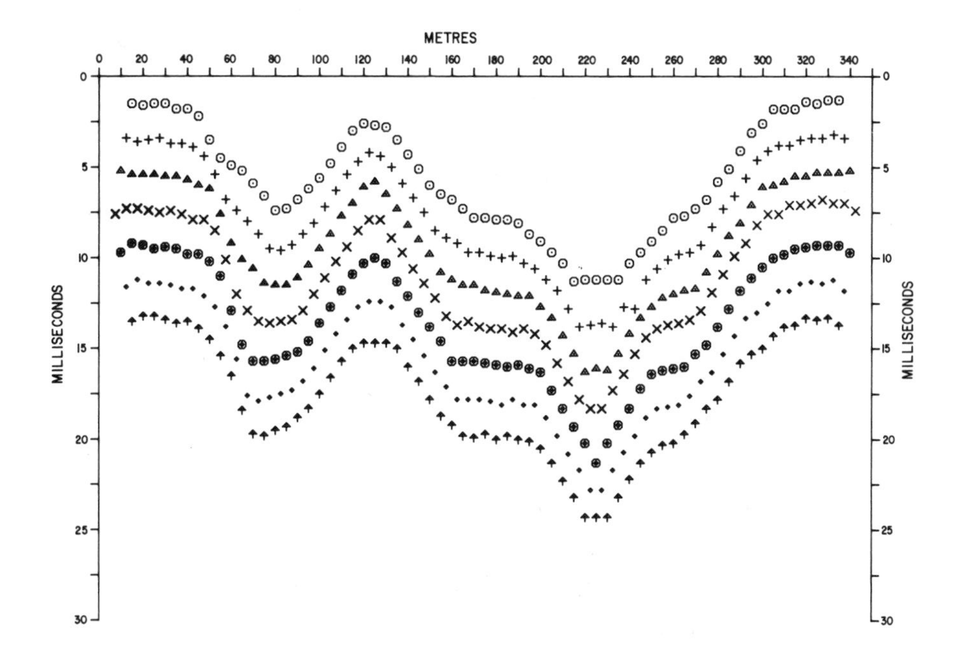

FIG. 7. Time-depths for XY-values from 0 to 30 m, derived from the traveltime data in Figure 5.

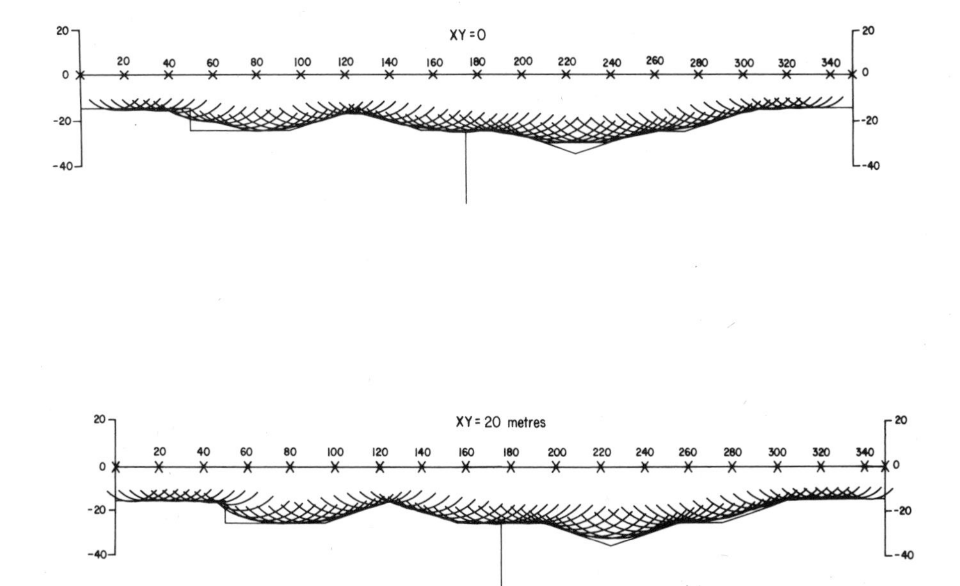

FIG. 8. Depth sections calculated from time-depths using 0 and 20-m XY-values.

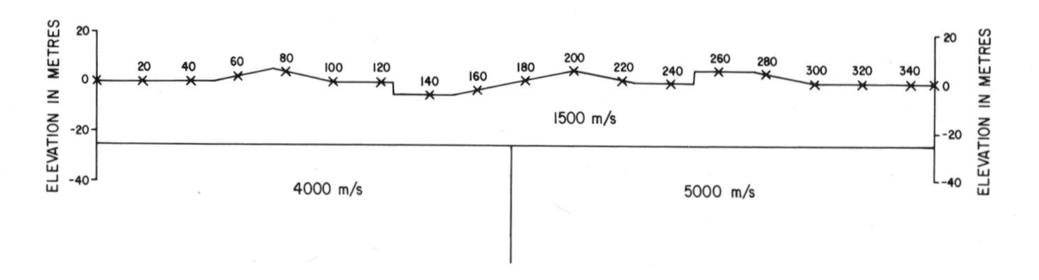

FIG. 9. Model with a very irregular ground surface and a plane horizontal refractor. The vertical and horizontal scales are equal.

For other XY-values, the points are scattered about the theoretically correct straight lines. In fact, for zero XY, the real change in the refractor velocity cannot be unambiguously recognized since it is obscured by numerous fictitious velocity changes. The change in refractor velocity of 1000 m/sec in this model is large and, in most situations, it would be geologically significant.

Time-depths for XY-values from 0 to 30 m are plotted in Figure 7. Depth sections calculated with 0 and 20 m XY-values are shown in Figure 8. The depth section for $XY = 20$ m is very close to the original model. The improved definition of the refractor surface compared with that for zero XY is obvious, particularly for the depression at 225 m and for the fault at 50 m.

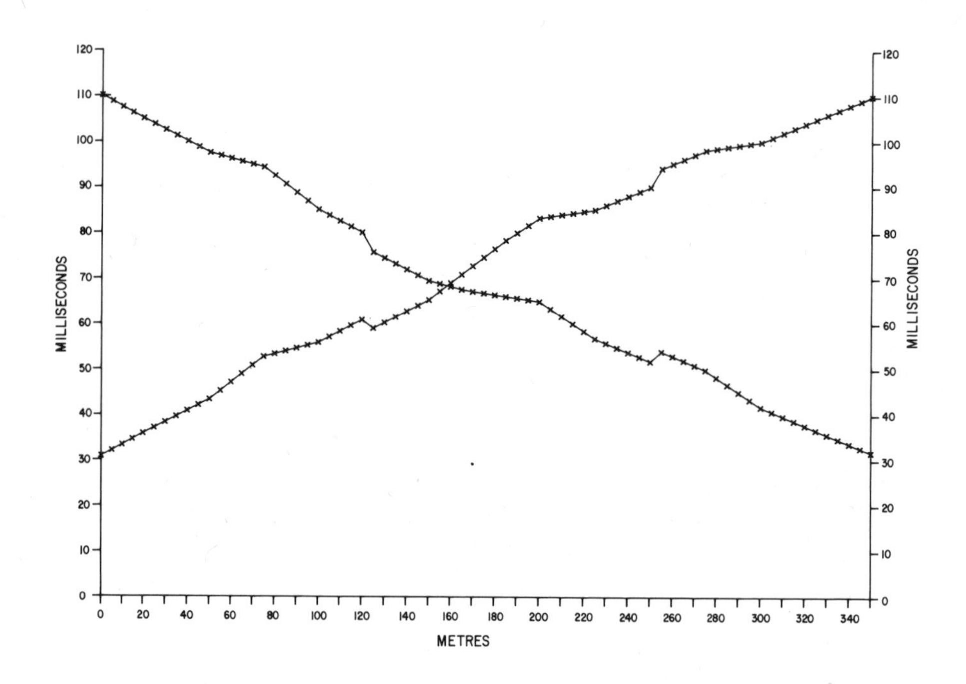

FIG. 10. Traveltime curves derived from the model in Figure 9.

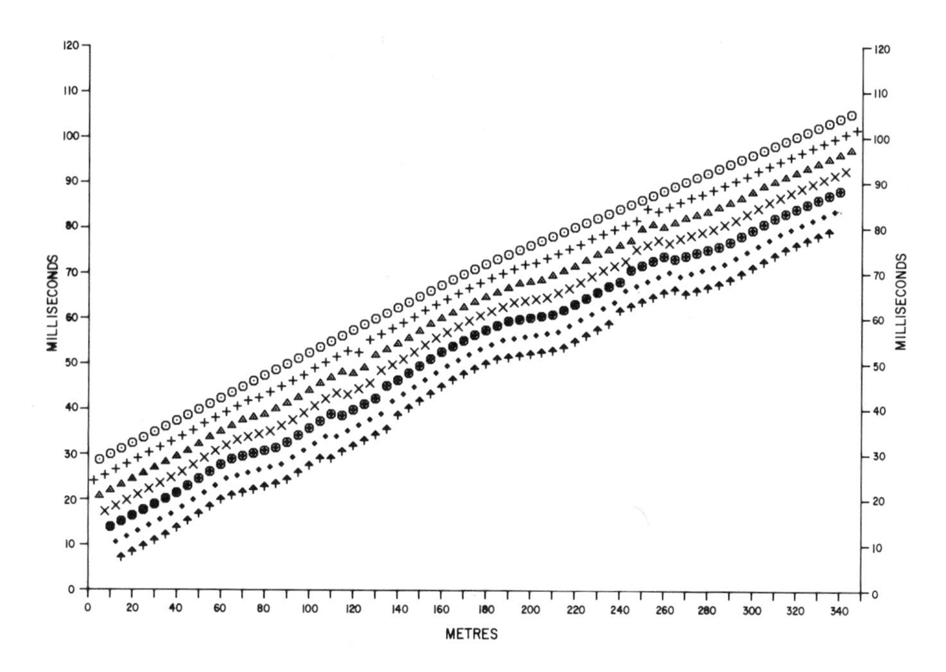

FIG. 11. Velocity analysis functions for XY-values from 0 to 30 m, derived from the traveltime data in Figure 10.

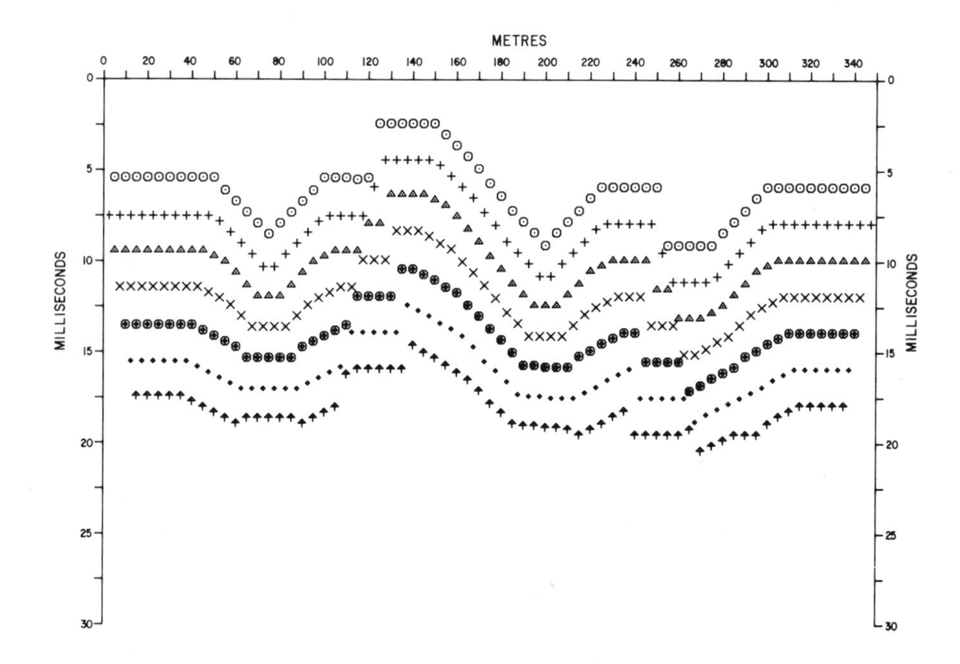

FIG. 12. Time-depths for XY-values from 0 to 30 m, derived from the traveltime data in Figure 10.

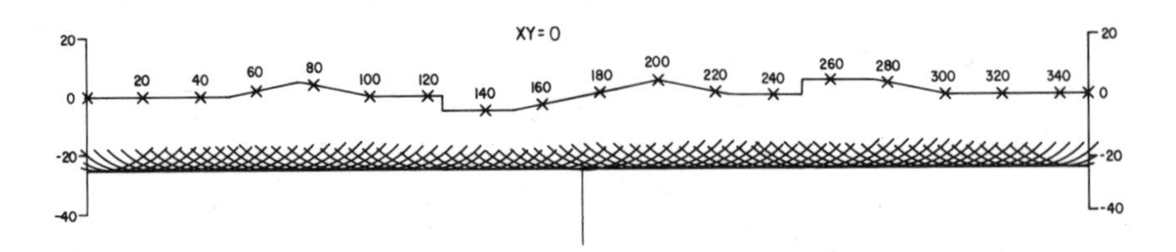

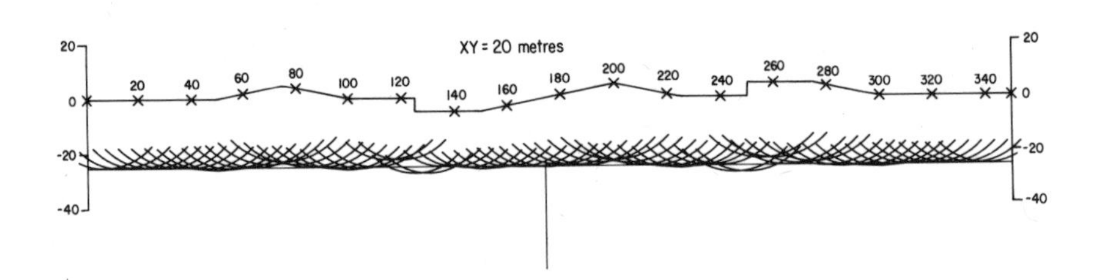

FIG. 13. Depth sections calculated from time-depths using 0 and 20-m XY-values.

Irregular ground surface

Another common condition in refraction work is irregular topography. The model shown in Figure 9 is the reverse of the previous model and has a very irregular ground surface and a plane horizontal refractor surface. Traveltime data for this model are listed in Appendix A, and time-distance plots are shown in Figure 10.

From the velocity analysis data in Figure 11, it can be seen that a zero XY-value provides the set of points which most closely approximate two straight lines. Time-depths for this model are plotted in Figure 12.

In the depth calculations for XY-values of 0 and 20 m in Figure 13, the irregular topography is accommodated with zero XY and the horizontal refractor is defined. At $XY = 20$ m, the irregular topography adversely affects the definition of the refractor. This is also true for other values of XY, if depths are plotted from the surface.

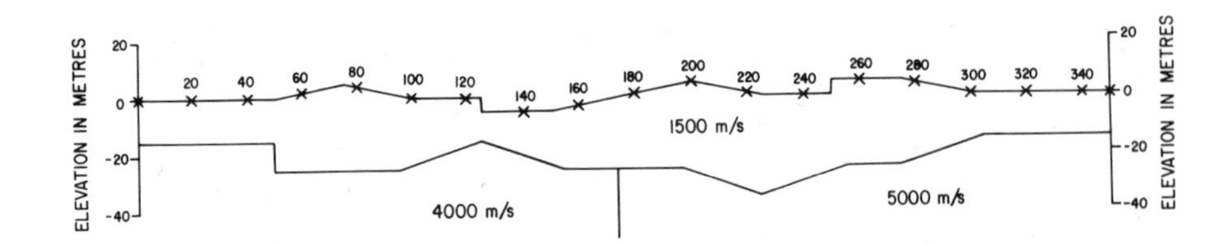

FIG. 14. Model with irregular ground and refractor surfaces. The vertical and horizontal scales are equal.

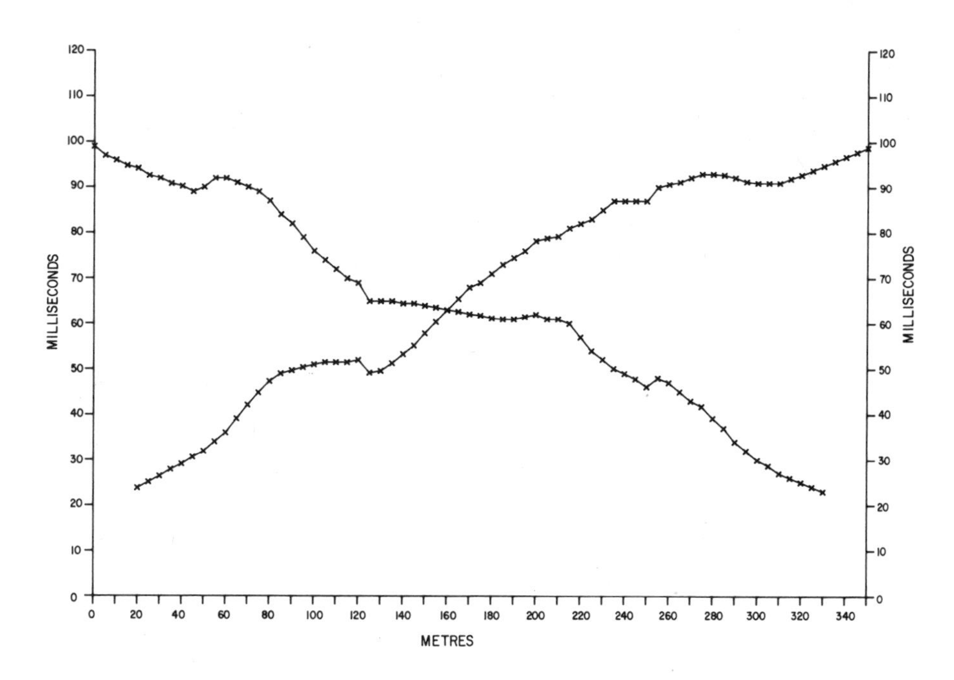

FIG. 15. Traveltime curves derived from the model in Figure 14.

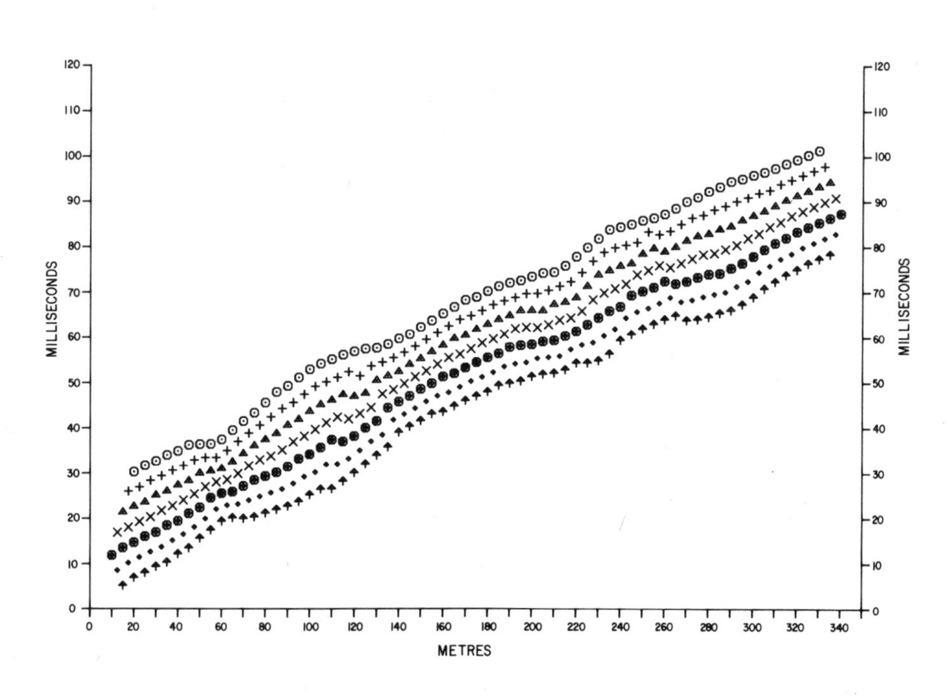

FIG. 16. Velocity analysis functions for XY-values from 0 to 30 m derived from the traveltime data in Figure 15.

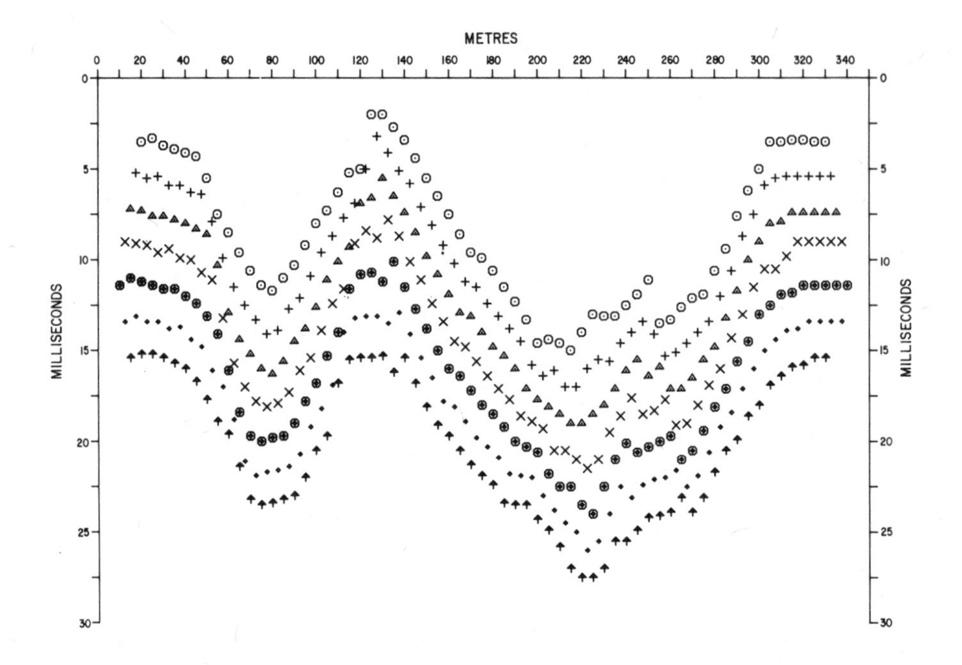

FIG. 17. Time-depths for XY-values from 0 to 30 m derived from the traveltime data in Figure 15.

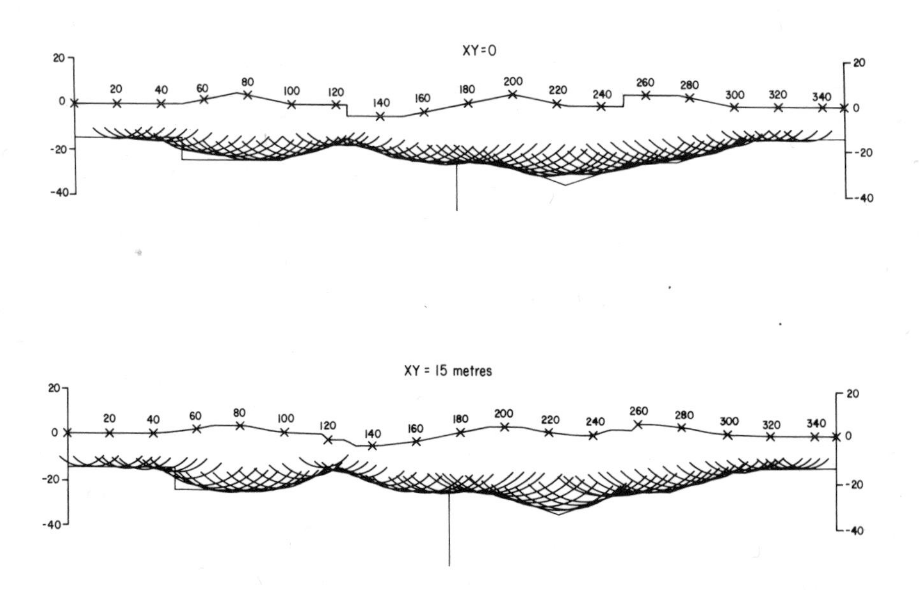

FIG. 18. Depth sections calculated from time-depths using 0 and 15-m XY-values. The ground surface shown in the 15-m case is the plotting surface, not the true surface.

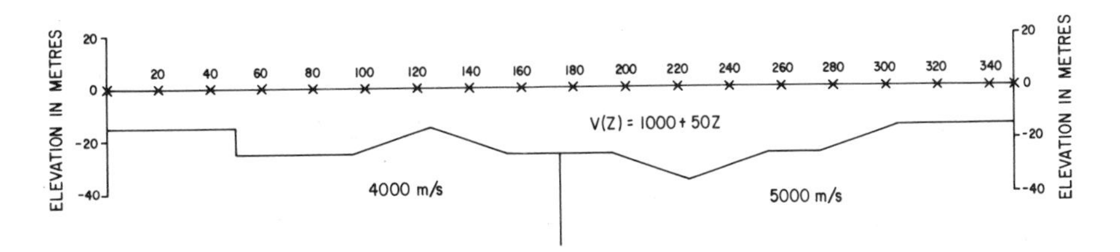

FIG. 19. Model with a plane horizontal ground surface, a highly irregular refractor, and a surface layer with a linear increase of velocity with depth. The vertical and horizontal scales are equal.

Irregular ground and refractor surfaces

The models shown in Figures 4 and 9 have been combined to form the model shown in Figure 14. Time-distance plots are shown in Figure 15, and traveltimes for this synthetic model are listed in Appendix A.

While no set of velocity analysis data shows two distinct straight line segments in Figure 16, the data for a 15-m XY-value are judged the best. Time-depths for a range of XY-values from 0 to 30 m are plotted in Figure 17.

The effect of the irregular ground surface in definition of the refractor can be minimized by plotting the depth at G from an elevation which is the mean of the elevations of X and Y. This procedure results in the requirements for plane layering being satisfied. For the special case of zero XY, this procedure is identical to that of Hawkins (1961, p. 810).

The depth sections in Figure 18 are comparable to those in Figure 8—the elevations shown are the mean of X and Y.

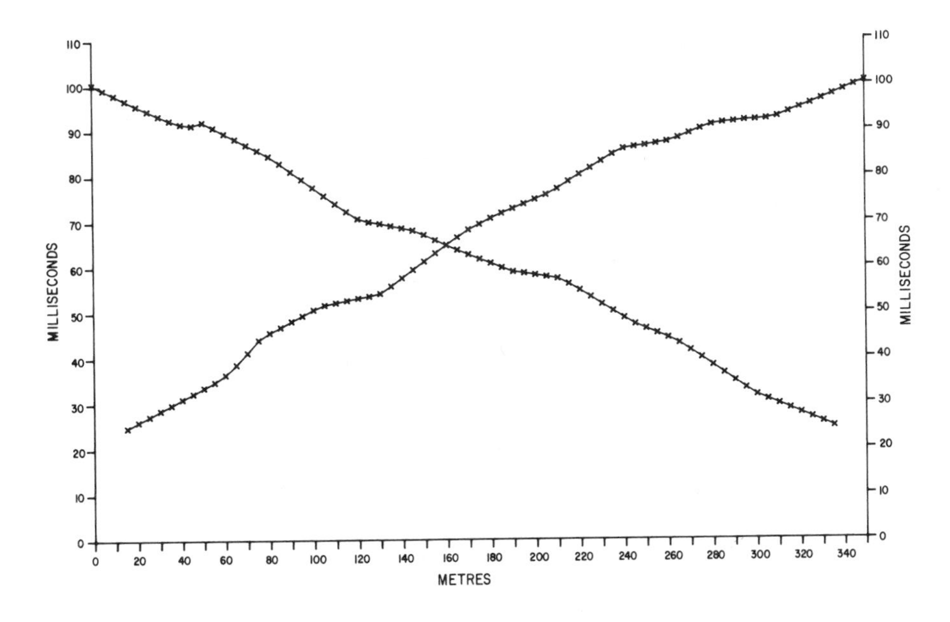

FIG. 20. Traveltime curves derived from the model in Figure 19.

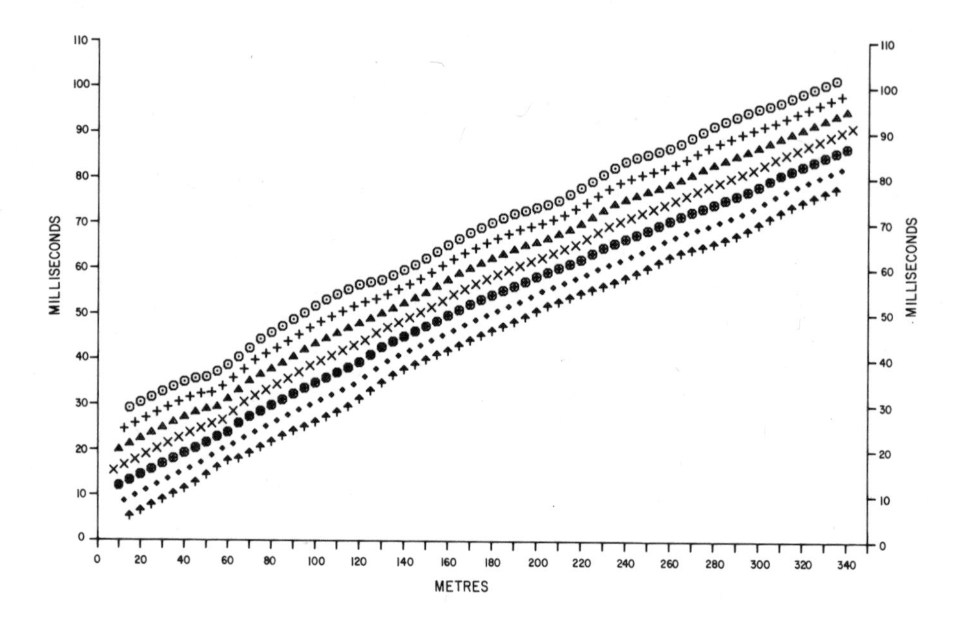

FIG. 21. Velocity analysis functions for XY-values from 0 to 30 m, derived from the traveltime data in Figure 20.

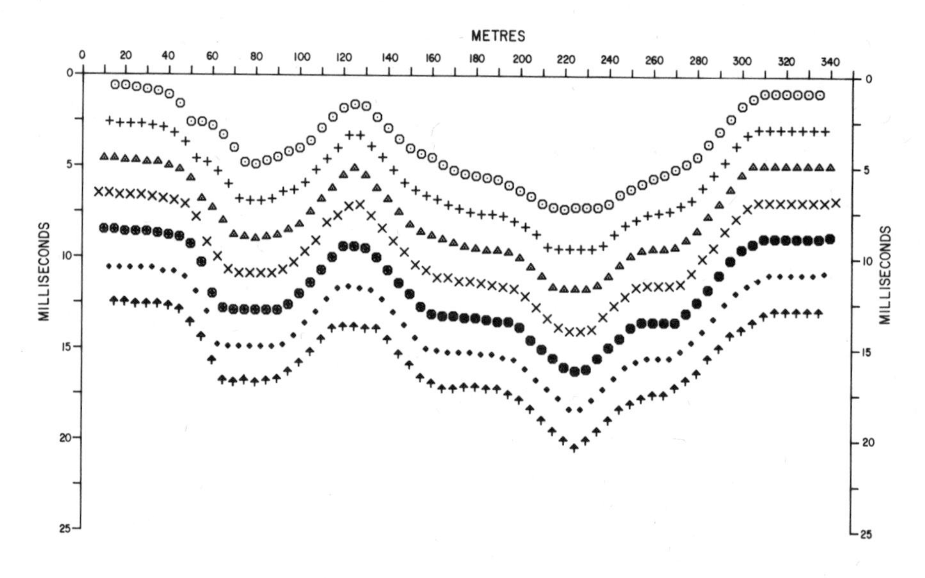

FIG. 22. Time-depths of XY-values from 0 to 30 m, derived from the traveltime data in Figure 20.

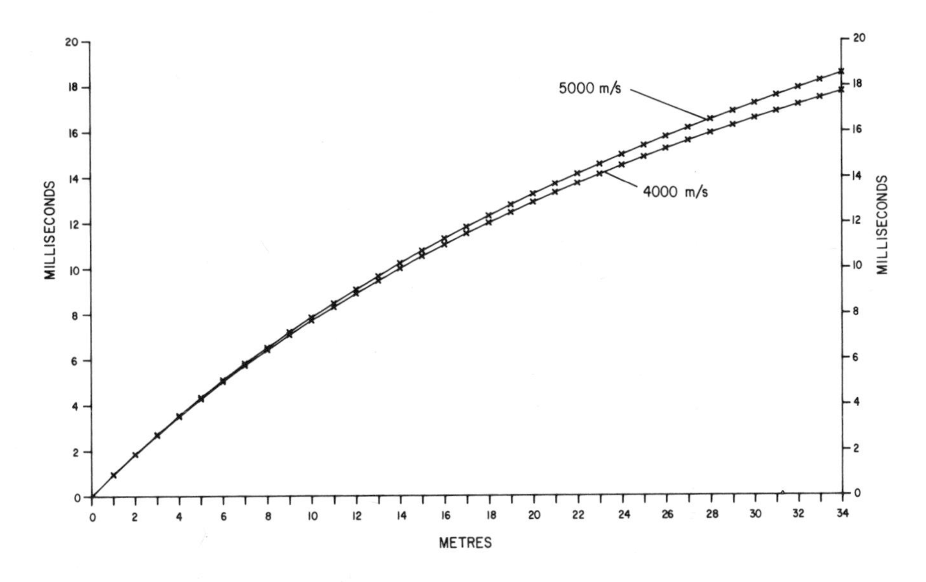

FIG. 23. Time-depths versus thicknesses for the linear increase of velocity with depth for two refractor velocities.

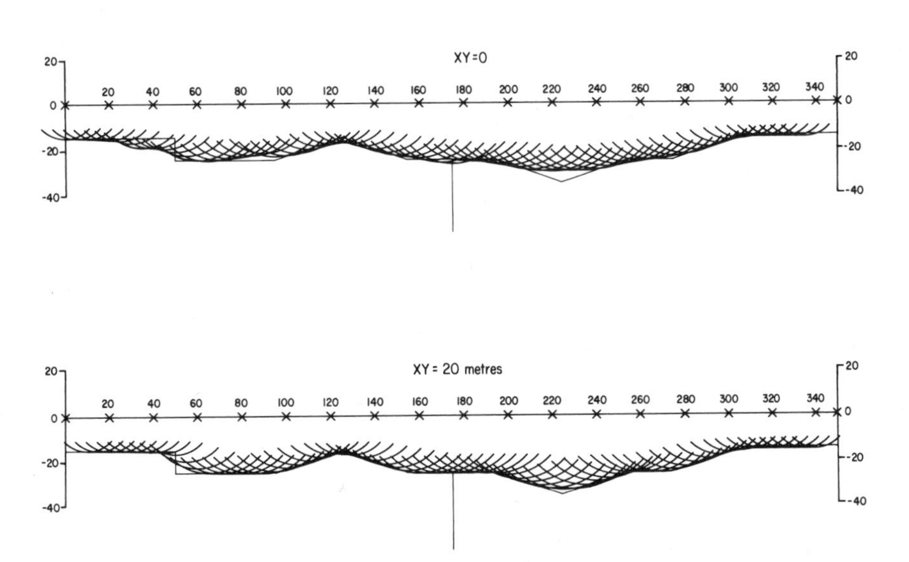

FIG. 24. Depth sections calculated from time-depth using 0 and 20-m XY-values.

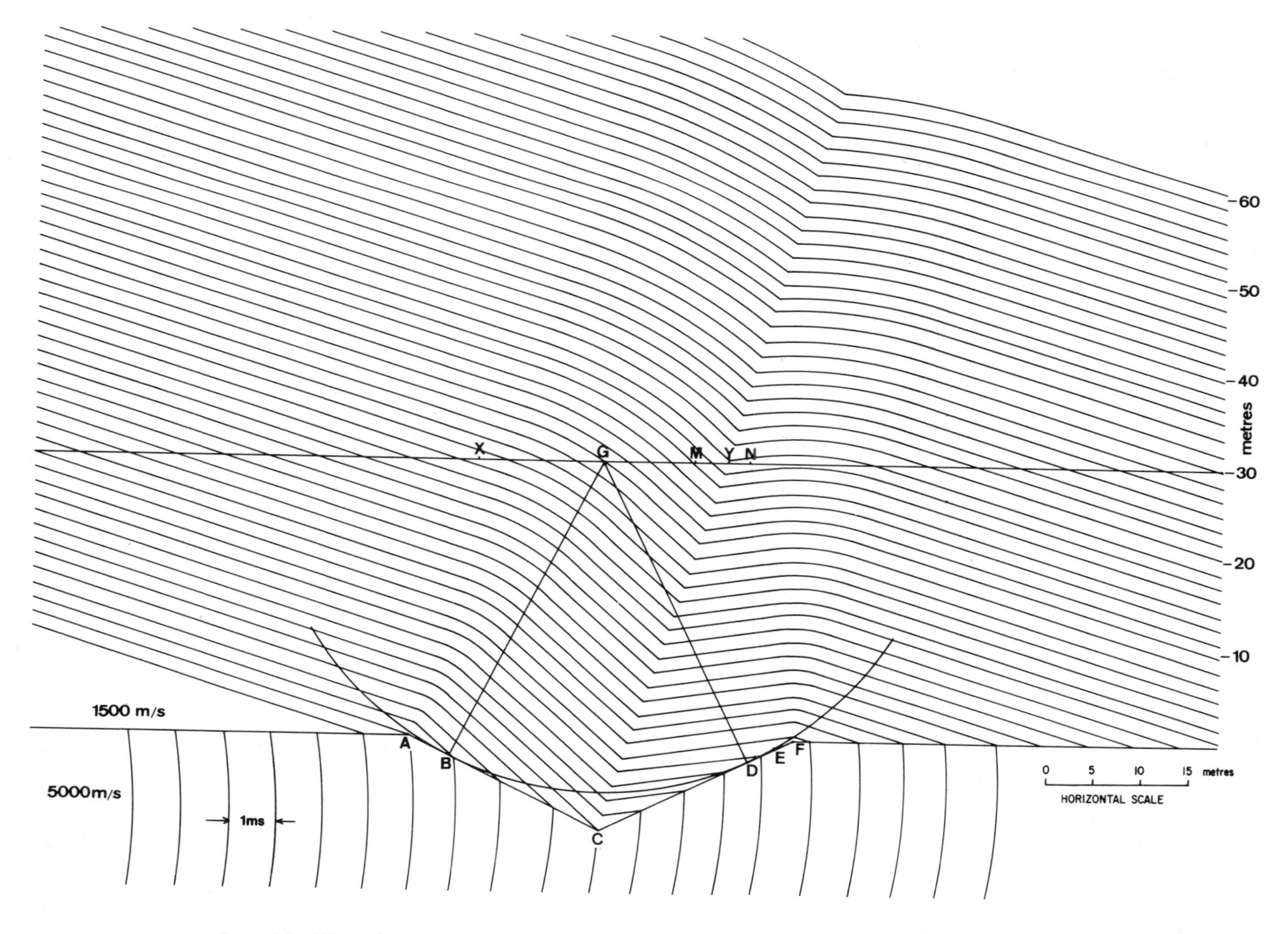

FIG. 25. Wavefront construction to demonstrate the smoothing of wavefronts resulting from increasing depth of refractor.

Continuous change of velocity with depth

In many areas, the velocity in clastic sediments increases continuously with depth because of the effects of compaction. Figure 19 shows a model with the same refractor horizon as in Figure 4, but now the velocity in the surface layer increases linearly with depth. The velocity function is

$$V(z) = 1000 + 50z \text{ m/sec.}$$

This velocity function results in a more rapid increase in velocity than is usually encountered in practice (cf., Dobrin, 1976, p. 311) and therefore provides an extreme test of the GRM. Traveltimes for this model are also listed in Appendix A, and a plot of the time-distance curves is shown in Figure 20.

The velocity analysis data in Figure 21 for an XY-value of 20 m fall very close to two straight lines, which provide velocities within 1 percent of the correct values. Time-depths are plotted in Figure 22.

Equation (19) has been evaluated and the results are summarized in Figure 23. Again, the depth section calculated from the time-depths using the optimum XY-value of 20 m is closer to the model than for other XY-values, e.g., $XY = 0\ m$ (Figure 24).

Detail of refractor definition

The preceding examples indicate that the GRM can provide very accurate definition of refractors. The definition is in fact limited by the information contained within the wavefronts originating from the refractor.

This can be shown with the aid of the model in Figure 25, where first-arrival wavefronts have been constructed for a horizontal refractor with a small steep-sided depression that extends 10 m below the refractor. It is assumed that first-arrival times can be determined for depths as great as 70 m to the horizontal surface of the refractor.

For refractor depths from 0 to 36 m, some first arrivals are obtained from the sloping surfaces of depression. However, first arrivals are not obtained from the central portion of the depression, because the arrivals from the updip side of the depression overtake arrivals from the downdip side. This can be seen for a refractor depth of 30 m in Figure 25. Refracted arrivals for the interval MN on the surface originate from the intervals AB and EF on the refractor. There are no first arrivals from the interval BCE on the refractor.

The updip and downdip wavefronts intersect and have a common arrival time which has the maximum observable delay associated with the depression. These wavefronts with the maximum delay in arrival time originate from points on the refractor, where the depths, measured perpendicular to the refractor, are maximum.

In Figure 25, for a refractor depth of 30 m, the maximum delay is observed at Y. The arrival at Y originates from the point B on the refractor, because BY is normal to the wavefronts. The distance BG is the maximum radius of any circle, which is centered on the line XGY tangential to AC and/or CF, and does not intersect AC or CF. This circle would represent the refractor surface as determined by the GRM.

Table 3. Depths of depression calculated by the GRM for increasing depths to the horizontal refractor surface.

Depth to horizontal refractor surface (m)	Maximum generalized time-depth (msec)	Calculated depth (m)	Maximum perpendicular from surface to refractor (m)
10	11¼	17½	18
20	17¼	27	27½
30	22¾	35¾	36½
40	28¼	44½	45
50	34¼	54	54½
60	39¾	62½	63
70	46	72½	72

For refractor depths greater than 36 m, the only first arrivals from the structure are diffractions from the edge of the depression.

In Table 3, maximum generalized time-depths and depths derived from these values are presented. Also shown is the maximum perpendicular distance from the surface to the refractor.

It can be seen that the depths calculated from the maximum generalized time-depths are, to sufficient accuracy, equal to the maximum perpendicular distance from the surface to the refractor. From this, the following conclusions can be made.

First, the GRM has extracted the maximum detail possible from the traveltime data. The limits of definition are a function of the refraction method itself, and not the GRM.

Second, the depth profile produced represents the *shallowest* depth of a refractor which can produce the observed traveltime data. This depth section will be termed the minimum equivalent refractor.

Third, the method of specifying depths of perpendicular thicknesses, as selected in chapter 2, is appropriate and results in refractor definition commensurate with the maximum detail provided by the refraction method.

Chapter 6
Selection of *XY*-values

Small-scale irregularities

Small-scale surface irregularities are defined as variations in thickness and/or velocity of near-surface layers (Gardner, 1967, p. 344) that extend over only a few geophone intervals at most. These irregularities are significant enough to invalidate interpolation between shotpoints, but are not sufficiently common to justify a field program which fully maps the surface layers.

Most interpretation methods aim to define an irregular refractor which is usually assumed to lie below uniform surface layers. However, the accuracy in defining deep refractors often depends upon the recognition and definition of small-scale surface irregularities. If the time anomalies caused by the surface irregularities are assigned to deeper layers, then the computed depths can be quite inaccurate, and the inferred refractor irregularities may not even exist. Furthermore, if XY spacings greater than zero are used in defining deeper layers, then depth anomalies computed from the time anomalies may be migrated away from their sources.

Because the surface layers usually have low seismic velocities, any variations in thicknesses or velocities produce time anomalies which can be many times larger than the anomalies produced by the same variations in layers nearer the refractor.

For example, at G in the model shown in Figure 3, a change of 1.2 m in thickness of the surface layer results in a variation of 1 msec in the arrival time. However, a change of 3.2 m in the thickness of the third layer is required for the same variation in the arrival time.

This effect can be even more accentuated in areas where the surface layers are dry and not compacted and are soils, alluvium, or weathered rock; in these instances, the seismic velocities can be less than 300 m/sec. In engineering surveys, it is not uncommon for sets of erratic traveltime curves to be caused by near-surface irregularities rather than variations in bedrock parameters.

There are three methods for recognizing these irregularities. The most accurate method is to obtain forward and reverse arrivals refracted from the surface layers, and hence define thickness and velocity variations with the GRM.

The usual method is by examination of the traveltime curves. Anomalies will only occur on those arrivals which include a traveltime through the irregularity. Therefore, anomalies will occur on traveltime curves which have arrivals from refractors as deep as or deeper than the irregularity. Hence, the source is the refractor represented by the segment of the traveltime curves on which the anomalies first occur.

Furthermore, the separation of the anomalies can assist. For surface irregularities, the anomalies will occur at the same geophone location on the forward and reverse traveltime from any refractor. However, for deeper sources, there will be some separation.

The third method is a refinement of the second method, which can be impractical when there are irregularities in both the surface and the refractor. The method requires the presentation of generalized time-depths for several XY-values.

For a near-surface inhomogeneity, arrival time anomalies will be observed on both forward and reverse times recorded at that location. For zero XY, an anomaly in the time-depth, equal to the anomaly in the arrival times, will be observed. For other XY-values, the time-depth anomaly will be halved and it will be observed in time-depths at $1/2$ XY on either side of the source of the anomaly.

This effect can be recognized in the generalized time-depths presented in Figure 12 for the model shown in Figure 9. The step in the topography at 250 m can be considered to be near-surface irregularity. For zero XY, the time-depth at 255 m is approximately 3 msec greater than that at 250 m. For other XY-values, the increase is about 1.5 msec and occurs at $1/2$ XY on either side.

For deeper irregularities, the maximum anomaly will be observed when the forward and reverse raypaths pass through the irregularity. For other XY separations, an anomaly at half the maximum will be located on each side. Refractor irregularities are a special case where these anomalies are maximized for accurate refractor definition.

This effect is demonstrated by the depression at 225 m in the example in Figure 4. For a 20-m XY-value, the time-depth at 225 m in Figure 7 is approximately 1 msec greater than the mean of the two adjacent values. For other XY-values, the increase is approximately 0.5 msec and occurs at 215 and 235 m for zero XY and at 220 and 230 m for 10-m and 30-m XY-values.

In Figure 14, both surface and refractor irregularities are present. However, it is still possible to recognize and separate both the surface irregularity at 255 m and the refractor irregularity at 225 m on the time-depth diagram (Figure 17). The separation of the time anomalies caused by these two features on the traveltime curves is quite difficult (Figure 15).

Determination of an optimum *XY*-value

There are several methods for determination of an optimum XY-value. First, when horizontal layering occurs, it can be very readily shown that when the forward and reverse rays emerge from the refractor at the same point,

$$XY = \sum_{j=1}^{n-1} 2\, Z_{jG} \tan i_{jn}. \tag{20}$$

Equation (20) can also be used for dipping layers, provided

$$|\theta_j - \theta_k| < 20°.$$

This can be shown with the example in Figure 3. In Table 4, the XY-value obtained by graphical construction of the raypaths and an XY-value calculated with equation (20) are presented.

It is clear that even with such an extreme model as shown in Figure 3,

Table 4. XY values determined by equation 20 and by graphical construction of raypaths.

Refractor velocity (m/sec)	Observed XY (m)	$\sum_{j=1}^{n-1} 2\, Z_{jG} \tan i_{jn}$ (m)	Error (percent)
2500	102	100.5	1.5
4000	79	82.7	4.5

the XY-value obtained with equation (20) is sufficiently accurate. Thus an XY can be calculated if the thickness of overburden and velocity distribution above the refractor are known from, for example, a continuous velocity log of a nearby borehole. Furthermore, it permits calculation of an expected XY-value for a given seismic depth section.

Second, the separation of distinctive features on the traveltime curves of forward and reverse shots can indicate a suitable value [see also Woolley et al, 1967, p. 280, (f)]. For example, in Figure 5, the change in slope at 235 m on the forward traveltime curve corresponds with the change in slope at 215 m for the reverse traveltime curve, and indicates that 20 m would be a suitable value.

Third, a value can be inferred from inspection of the velocity analysis and time-depth calculations for several XY-values. The XY-value for which the velocity analysis curves are the simplest, and the time-depths show the most detail, corresponds to the optimum value.

The time-depths for XY-values other than the optimum tend to be less detailed, or smoothed, compared with the time-depths for the optimum XY. However, small-scale irregularities from other layers, usually from near the surface, can result in time-depths for nonoptimum XY-values being more irregular than time-depths for the optimum XY. Therefore, the sources of irregularities of small lateral extent must be identified; otherwise the time-depths affected by these irregularities may be chosen as the most detailed, and an incorrect XY-value may be selected.

One approach has been to attempt recognition of simple steps or faults in the refractor. The point at which the time-depths begin to rise, or fall, corresponding to the edge of the step is located. For the optimum XY, this point is closest to the true position of the step. With nonoptimum XY-values, the time-depths are smoothed, so the rise or fall in time-depths occurs farther away from the edge of the step.

The velocity analysis curves can also be an excellent indicator of the optimum XY. Figure 26, which is taken from Figure 6, is a schematic representation of velocity analysis function for a step model. At the optimum XY, the velocity analysis function is a straight line. For other XY-values, the velocity analysis curves depart from straight lines for a short interval, which is a function of the optimum XY-value on either side of the step and of the actual XY-value used. In the example shown, for a shotpoint on the left-hand side of the section, the curves converge toward the optimum XY-line, while for curves plotted for the reverse shotpoint, divergence will occur.

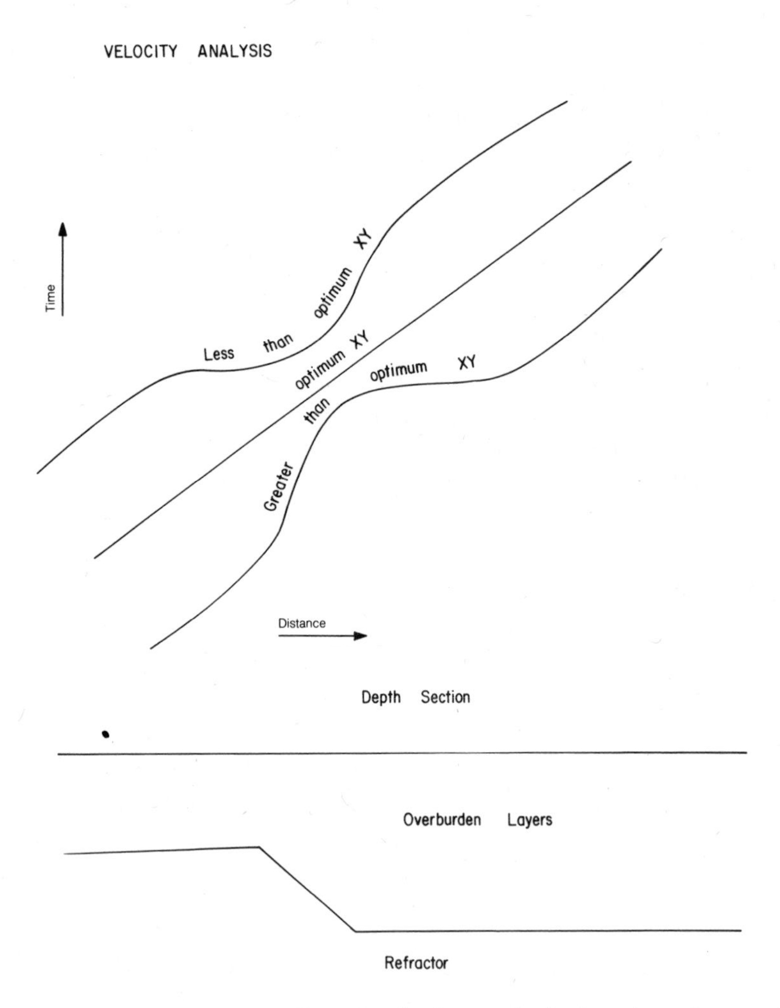

FIG. 26. Schematic representation of velocity analysis functions for a refractor with a step in the depth.

Fourth, an XY-value is the distance at which the critical reflection occurs (Grant and West, 1965, p. 108; Layat, 1967, p. 179). At the critical reflection, marked increases in reflection and refraction amplitudes occur.

Summary

The determination of optimum XY-values is potentially the most confusing aspect of the GRM. On one hand, examples which have large variations in dip and depths indicate that the selection of an XY-value is not critical in depth calculations. On the other hand, as will be shown in the next chapter, determination of accurate XY-values is essential for the detection of unrecorded layers.

For normal application of the method, it is generally prudent to compute and plot GRM parameters for XY-values ranging from zero to in excess of the expected optimum, rather than for just one value (be it zero or the optimum). This procedure permits determination of an appropriate value, if the refractor is sufficiently irregular. However, if an optimum value cannot be determined, then the full presentation verifies that the GRM parameters are a result of a featureless refractor and not an inappropriate XY-value, and that any convenient XY-value, e.g., zero, can be used for depth calculations. Full presentation of GRM parameters also permits recognition and definition of near-surface irregularities, an essential operation, irrespective of which XY-value is used.

Accurate determination of XY-values from time-depth and velocity analysis plots is difficult and is made more so by the smoothing inherent in first arrival data. It is possible that more accurate traveltime data, as well as statistical methods, would yield considerable improvement in this determination.

Chapter 7
Undetected layers

The blind zone

In elementary treatments of the seismic refraction method, each layer is assumed to have a uniform velocity and to be bounded by plane surfaces. For these simple models, the field data consist of traveltime curves, each segment of which represents a different refractor. A depth section is readily obtained by assigning a refractor to each segment and employing one of the standard formulas.

When the geologic situation deviates from the plane interface model, then the more advanced interpretation methods are used to map the irregular refractor surface. However, while more advanced routines recognize the existence of irregular refractors, it is still commonly assumed that the velocity stratification can be unambiguously inferred from the traveltime curves. This assumption constitutes probably the most serious shortcoming of the refraction method (Hagedoorn, 1959, p. 164–166; McPhail, 1967, p. 260).

In many cases, improved field procedures are sufficient to resolve the inherent ambiguity of single traveltime curves. To separate the effects of changing refractor dip from the recording of other refractors, at least four shots—two from different locations at each end of the geophone spread—are necessary (see Hawkins, 1961, p. 810). For example, more thorough field programs might have resulted in Duguid's (1968) paper being less open to discussion (Sendlein, 1968).

Unfortunately, an increase in the number of shotpoints is not a solution to all problems of ambiguity. One example is the hidden layer where energy from a refractor of higher seismic velocity arrives at the surface before energy from an overlying refractor. Since only first arrivals are used in most seismic refraction studies, thicknesses and velocities of the overlying refractor cannot be calculated, and, consequently, depths to lower refractors are in error (Soske, 1959).

The *hidden layer* or *masked layer* thickness can vary between zero and a maximum theoretical thickness, which is termed the *blind zone* (Hawkins and Maggs, 1961, p. 526). In theory, when the layer exceeds the blind zone thickness, it occurs as a first event and is therefore detected. However, since arrivals at several geophone locations are usually required before the existence of another layer can be stated with confidence, the actual thickness of hidden layers can in fact exceed the blind zone thickness.

Maillet and Bazerque (1931, p. 314) presented a nomogram for the thickness of the blind zone in terms of the depth to the top of the blind zone. A more convenient nomogram, in which the blind zone is stated in terms of the depth to the recorded refractor calculated without considering a velocity change within the blind zone, is presented by Hawkins and Maggs (1961). Their treatment is similar to those of Shima (1957) and Berry (1971). Further extension of the blind zone to three- and four-layer cases, with an arbitrary shot depth in the top layer and for media in which the velocity increases linearly with depth, is presented by Puzyrev (1972). The treatment of Merrick et al (1978) can accommodate any number of blind zones within a multilayer sequence of horizontal or dipping layers.

The nomograms of Hawkins and Maggs (1961) have been criticized for only presenting the maximum thickness of the hidden layer, i.e., the blind zone (Green, 1962). However, as stated in Hawkins (1962), this is the only parameter which can be determined in the absence of any other data.

The blind zone is more than a measure of the maximum error in depth calculations caused by hidden layers. It represents the zone in each layer where the velocity distribution, which is determined in the upper part of the layer, is extrapolated (Hagedoorn, 1955, p. 329–332). The blind zone is a necessary consequence of the basic characteristic of the refraction method, in which arrivals from a deeper layer overtake those from a shallower layer, or part of a shallower layer. *Therefore, every layer has a blind zone, but not necessarily a hidden layer within the blind zone.*

The example of Hagedoorn (1955, p. 329–332) emphasizes the significance of the blind zone. Even when hidden layers are absent, the velocity distribution in the blind zone still cannot be accurately obtained by extrapolation from the upper part of the layer.

Second events

The use of second and later events to detect hidden layers has been advocated (Green, 1974, p. 276). However, the reduced accuracy in measuring arrival times of these later events often limits their usefulness, particularly in the vicinity of the cross-over distance where several high-amplitude wavelets can interfere. Sometimes, second events simply do not occur (Dobrin, 1976, p. 317).

Particular care is necessary to distinguish second events that are representative of hidden layers with uniform velocities from triple-valued traveltime curves that are representative of layers with velocity gradients (Slichter, 1932, p. 280; Goguel, 1951, p. 20). For example, it is possible that the traveltime curve presented by Kaila and Narain (1970, p. 615) may be triple valued and so represents a section in which the velocity increases continuously with depth.

Furthermore, minor fluctuations in velocity with depth can produce a profusion of second events which frequently have much larger amplitudes than the first arrivals (Green and Steinhart, 1961, p. 587). Therefore, a problem can exist in determining whether second events can be assigned to a true hidden layer, a layer with a velocity gradient, or to minor variations in a complex velocity-depth profile.

Statistical evidence (Faust, 1951, 1953; Acheson, 1963) suggests that layers with velocities increasing with depth are more probable than constant-velocity layers. Berry (1971, p. 6467) shows that a series of layers with the velocity increasing with depth can explain a complete traveltime curve of linear segments with no later events. Also, continuous velocity logs indicate that minor fluctuations in velocity with depth are very common. If the Wiechert-Herglotz-Bateman integral is employed to detect hidden layers, then all loops in the traveltime curve must be used (Grant and West, 1965, p. 139–141), otherwise serious errors will result. Therefore, while second events have been used with some success (Dobrin, 1976, p. 315–318), their use in solving the hidden-layer problem is not as straightforward as has been suggested.

Velocity inversions

Another source of errors in refraction depth calculations is the velocity inversion problem (Domzalski, 1956, p. 153–155; Knox, 1967, p. 207–211). Critically refracted rays are not produced from the top of layers which have a lower seismic velocity than that of the overlying layer, unless the overlying layer is thin in relation to the wavelengths of the recorded waves and is therefore not an adequate refractor. Under certain circumstances, the existence of a velocity inversion can be inferred from time delays or "skips" on field records (Greenhalgh, 1977, p. 179). In general, depth calculations can be subject to unknown but often large errors because of this problem.

Detection of hidden layers and velocity inversions

Errors in depth calculations caused by hidden layers or velocity inversions can be avoided when drill holes with either lithological or velocity logs, or average velocities from seismic reflection surveys, are available.

However, when these data are not available, it may still be possible to recognize and define undetected layers through consideration of XY-values. There are two methods of obtaining XY-values. The first is from analysis of time-depth and velocity plots; the second is by computation from the depth section using equation (20). *If the depth section is to be consistent with the traveltime data, the computed and observed XY-values must agree.*

Furthermore, if the XY-values agree, then the depth to the refractor will be essentially correct, even though the thicknesses and velocities of intermediate layers may be in error. This follows from the next chapter where it is shown that accurate depths can be calculated using an average velocity derived from the XY-value, the time-depth, and the refractor velocity.

From equation (20), it can be seen that for a given layer, the XY-value increases when the thickness and/or the velocity of the layer increases. Therefore, if the observed XY-value is greater than the calculated value, then the higher velocity layers may be thicker, the layer velocities may be higher, or a blind zone may occur. Conversely, if the observed XY-value is less than the calculated value, then the low-velocity layers may be thicker, layer velocities may be lower, or a velocity inversion may occur.

Summary

Undetected layers are the most common source of errors in the majority of refraction interpretation methods, because deeper layers are defined with equations containing thicknesses and velocities of shallower layers. Even when all layers are detected, there is still a zone of uncertainty in each layer, the blind zone, in which the velocity is determined by extrapolation from the upper part of the layer. The blind zone, the zone of extrapolation, is an inescapable phenomenon of the refraction method and is an inevitable consequence of energy from deeper layers arriving at the surface before energy from shallower layers.

Therefore, it is recommended that the verification of the existence or absence of undetected layers, by correlation with drill hole or reflection data, or by comparison of observed and computed XY-values, be a routine step in all refraction interpretation methods.

Chapter 8
Average velocity

Definition

The use of an average velocity above the refractor permits depth calculations without defining all layers. It can also be useful in accommodating undetected layers, as discussed in the previous chapter. The method described below uses the optimum XY-value, but, unlike the methods of Hawkins [1961, equation (5)] and Woolley et al (1967, p. 279–280), a depth to the refractor is not required.

With the substitution of the horizontal-layer approximation, equation (12) becomes

$$t_G = \sum_{j=1}^{n-1} Z_{jG} \cos i'_{jn} / V'_j, \tag{21}$$

where

$$\sin i'_{jn} = V'_j / V'_n. \tag{22}$$

At the optimum XY separation, the forward and reverse rays emerge from near the same point on the refractor and

$$XY_{\text{optimum}} = \sum_{j=1}^{n-1} 2\, Z_{jG} \tan i_{jn}. \tag{23}$$

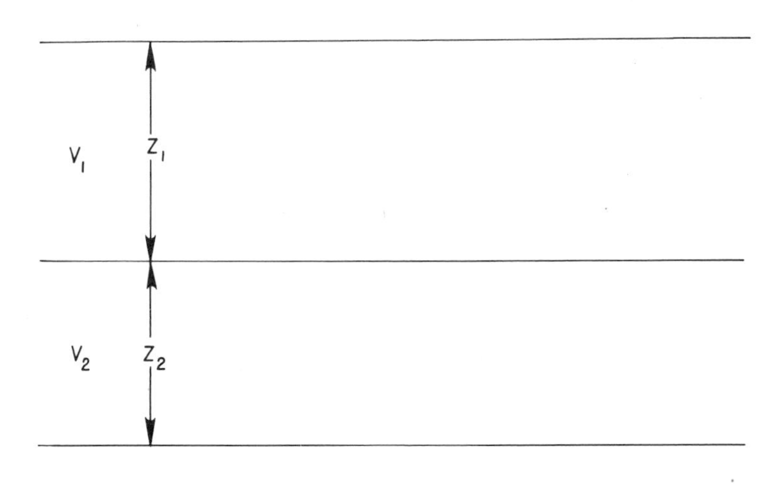

FIG. 27. Summary of symbols for the three-layer case.

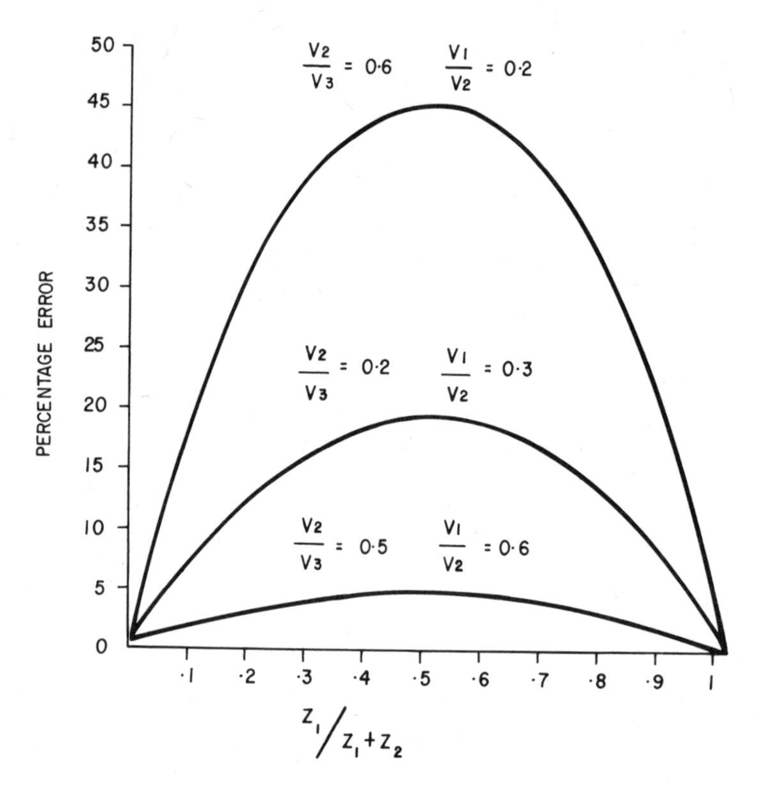

FIG. 28. Errors in depths calculated by the average velocity method, for three velocity contrasts. The maximum error occurs when the layers are of equal thickness.

An angle $\bar{\imath}$ and an average velocity $\bar{V}$ can be defined such that

$$\sin \bar{\imath} = \bar{V}/V'_n, \tag{24}$$

$$t_G = \frac{\cos \bar{\imath}}{\bar{V}} \sum_{j=1}^{n-1} Z_{jG}, \tag{25}$$

and

$$XY = 2 \tan \bar{\imath} \sum_{j=1}^{n-1} Z_{jG}. \tag{26}$$

These equations can be combined to form the following expression

$$\bar{V} = [V'^2_n \, XY/(XY + 2\, t_G \, V'_n)]^{1/2}. \tag{27}$$

For field examples, the calculations of time-depths using equation (10) and refractor velocities using equation (6) present few problems. Therefore, if an optimum XY-value can be determined, then an average velocity can be calculated with equation (27). The total thickness of all layers can then be computed by rearranging equation (25).

Errors for the three-layer case

An appreciation of the efficacy of equation (27) can be obtained by comparing depths calculated using the average velocity with the actual depths for a fully defined model. The model to be considered (Figure 27) has two layers above the refractor. Although all interfaces are plane and horizontal for ease in computation, the results are considered valid for dips up to 20 degrees, the limit of the GRM horizontal-layer approximations.

The total depth is calculated from equation (25), after an average velocity has been determined by substituting time-depth and XY-values into equation (27). Since there are no field data for this synthetic example, appropriate values of time-depth and XY must first be computed with equations (21) and (23).

Figure 28 shows the depth errors as a function of the proportion of the top layer to the total thickness for three velocity distributions. It can be seen that the maximum error for each velocity distribution occurs where each layer is of equal thickness.

In Figure 29, the maximum errors are presented for the complete range of velocities for the three-layer model. A maximum error of less than ten percent occurs when the seismic velocities of the overburden layers are similar (i.e.,

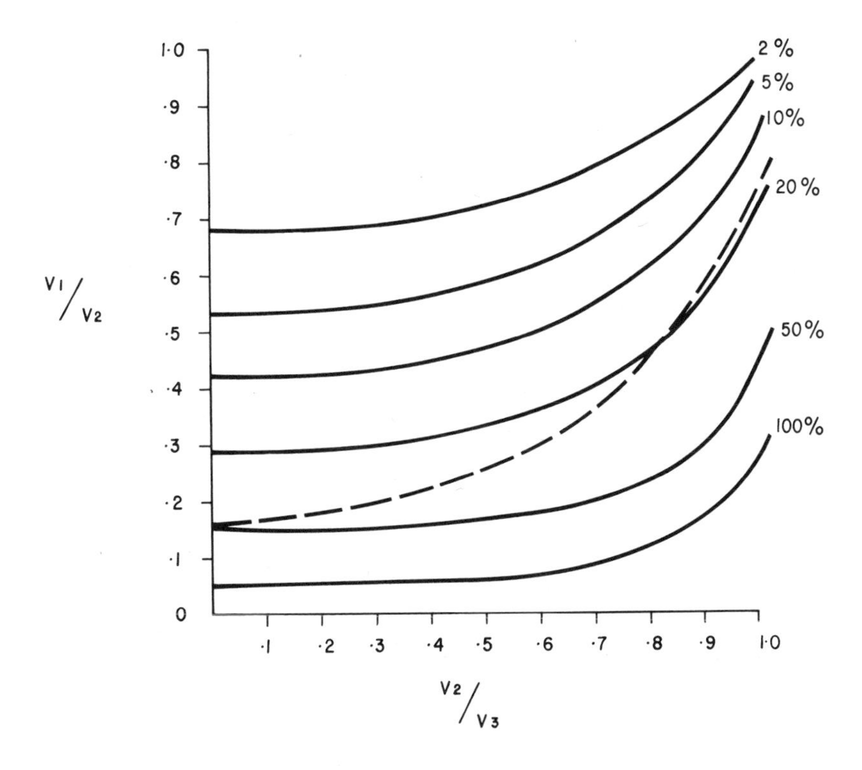

FIG. 29. Maximum errors in depths calculated by the average velocity method, for the complete range of velocity contrasts for the three-layer model. These errors are the same for a velocity inversion in the upper layers. In the area below the dashed line, the errors are greater than those in Figure 31 for the blind zone problem.

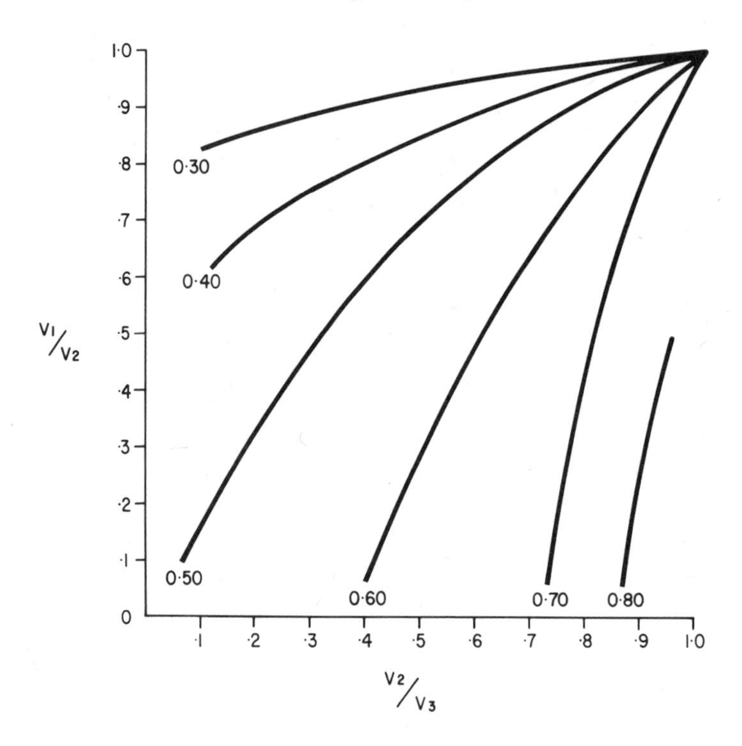

FIG. 30. Values of the ratio $Z_1/(Z_1 + Z_2)$ at which the blind zone occurs.

when $V_1/V_2 > 0.5$) and when there is a good contrast between the seismic velocities of these layers and that of the refractor (i.e., when $V_2/V_3 < 0.6$). These velocity contrasts could be expected in the majority of field cases to which the three-layer model is applicable.

Comparison with hidden-layer errors

The errors in depths calculated with the average velocity method are considerably less than those encountered when the second layer occurs as a hidden layer. The hidden layer exists when arrivals from that layer occur after arrivals from deeper layers (see chapter 7). The maximum thickness of the hidden layer (i.e., the blind zone) occurs when (Maillet and Bazerque, 1931, p. 313)

$$Z_1/Z_2 = \frac{\cos i_{23} \sin i_{12} (1 - \sin i_{12})}{\sin (i_{12} - i_{13}) + \cos i_{12} - \cos i_{13}}. \tag{28}$$

This ratio is present in Figure 30 for a complete range of velocity ratios.

If the hidden layer is not detected, then the maximum error that can occur in the depth calculations is

$$\frac{\cos i_{23} \sin i_{12} - \cos i_{13}}{\cos i_{13} (1 + Z_1/Z_2)}, \tag{29}$$

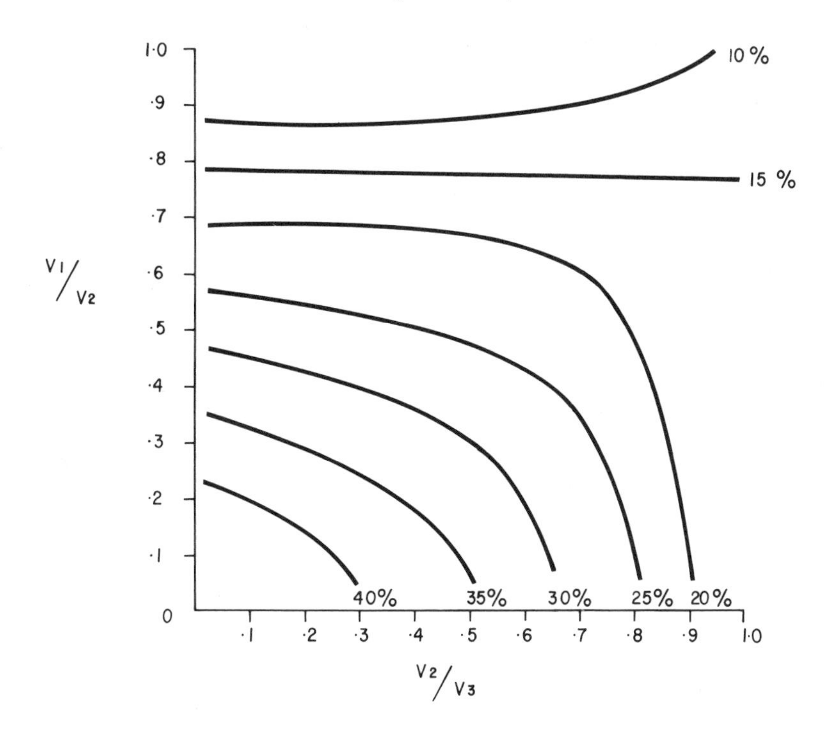

FIG. 31. Maximum errors in depth calculations, if the lower layer occurs within a blind zone and is not considered in depth calculations.

where Z_1/Z_2 takes the value appropriate for the blind zone. This error is shown in Figure 31.

For the velocity ranges $V_1/V_2 > 0.5$ and $V_2/V_3 < 0.6$ (apart from the trivial case when $V_1/V_2 > 0.9$), the maximum error which would result, if the blind zone were not considered in depth calculations, is greater than ten percent. In general, for each set of velocity ratios within the quoted range, these errors are four or more times larger than the errors for the average velocity method.

In the area below the dotted line on Figure 29, the blind zone errors are less than the average velocity method errors. However, the differences would not be as large if the Z_1/Z_2 ratio appropriate to the blind zone were used in the average velocity calculations. From Figures 29 and 30, it is apparent that the errors for the average velocity method would be reduced considerably.

Corrections for surface layers

The accuracy of depth determinations with the average velocity is adversely affected by large velocity contrasts in the layers above the refractor. It is usually the dry surface layer which has seismic velocities greatly different from the deeper layers, and, in general, it is usually possible to delineate these layers.

Therefore, it is possible to increase the accuracy of the average velocity by subtracting the contributions of the surface layers to the XY- and time-depth values, i.e.,

$$XY = XY_{\text{observed}} - 2 \sum_{j=1}^{k} Z_j \tan i_{jn}, \tag{30}$$

and

$$t_G = t_{G\text{ observed}} - \sum_{j=1}^{k} Z_j \cos i_{jn} / V_j, \tag{31}$$

where there are k surface layers. The average velocity then applies to the remaining layers.

Synthetic example

The synthetic example in chapter 5, with the continuous change of velocity with depth, provides a fairly severe test of the average velocity method. An optimum XY-value of 20 m was determined from the velocity analysis data. Also, an average time-depth value of 15 msec, which corresponds with a depth of about 25 m, will be used.

To facilitate depth calculations, equations (27) and (25) have been combined to form

$$\sum_j Z_{jG} = t_G \sqrt{\frac{V_n' XY}{2\bar{t}_G}}, \tag{32}$$

where $\bar{t}_G$ is the average time-depth value used in equation (27).

In Table 5, depths computed using equation (19) are listed and compared with those obtained with equation (32).

Table 5 also shows depths obtained by multiplying time-depths by 1.2. This probably represents the largest depth conversion factor which might be obtained by assuming that first arrivals represent a constant velocity in the upper layer. This assumption is not unreasonable because the first arrival data, obtained with equation (9.44) of Dobrin (1976, p. 310) are, within a millisecond or so, almost linear with distance. The arrival times at the distances indicated are:

Table 5. Comparison of depth errors obtained using equation 32 and by calculation assuming a constant velocity upper layer.

Location (m)	Time-depth (msec)	Depth by eq. 19 (m)	Depth by eq. 32 (m)	Error (percent)	Time-depth times 1.2 (m)	Error (percent)
15	10.5	15.0	17.1	14	12.6	−16
80	14.9	25.1	24.3	−3	17.9	−29
125	11.4	16.8	18.6	13	13.7	−19
225	18.2	33.1	33.2	—	21.8	−34
260	15.5	25.8	28.3	12	18.6	−28
330	10.9	15.0	19.9	25	13.1	−13

Distance in meters	0	5	10	15	20	25	30	35	40
Arrival time in milliseconds	0	5	10	15	19	24	28	32	35

It is clear that depths calculated using an average velocity are generally more accurate than those using a velocity obtained from uncritical acceptance of the observed traveltime data.

A further improvement in depth calculations can be obtained by approximating the upper layer with two layers. A seismic velocity of 1000 m/sec for the surface layer is suggested by the traveltime data. For the second layer, a seismic velocity of 1800 m/sec will be assumed, because this velocity is often representative of saturated unconsolidated alluvium or completely weathered bedrock. Also, an average seismic velocity of 4500 m/sec for the refractor will be used.

It is possible to form two linear equations after equations (21) and (23) with the unknown variables being the layer thicknesses. For the time-depth and optimum XY-value used previously, the thicknesses are about 5 m for the upper layer, and about 19 m for the lower layer. The following equation can then be formed, relating total depth with time-depths

$$\sum_j Z_{jG} \simeq 5 + 2\,(t_G - 5).$$

It is left to the reader to verify that this relationship results in a further improvement in depth estimates.

The example illustrates that use of an average velocity, or even simple consideration of the optimum XY-value, can result in improved depth estimates.

Summary

The determination of an average velocity using the XY-value in refraction interpretation can be considered to be analogous to the determination of root-mean-square (RMS) velocity from normal moveout measurements in reflection interpretation. As in the case of depth conversion of reflection times with RMS velocities, the average velocity permits accurate depth calculations, irrespective of whether all layers have been detected. Also, it overcomes the problem of small errors in depths to the upper layers resulting in large errors in depths to the deeper layers.

Accuracy of the average velocity determinations is primarily a function of the accuracy of the XY-value, because the calculation of time-depths and refractor velocities usually presents few problems.

Chapter 9
Reciprocal time

In the computation of time-depths and velocity analysis functions, the reciprocal time, the time from shotpoint to shotpoint, is required. However, of all the times determined in a seismic refraction profile, the reciprocal time is the most difficult to determine accurately. There are several reasons for this.

First, the reciprocal geophone is the most distant from the shotpoint. It receives the least amount of energy because head waves are attenuated approximately as the inverse square of the distance (Grant and West, 1965, p. 181). Furthermore, the earth acts as a low-pass filter, with the higher frequencies being attenuated more rapidly than lower frequencies (Attewell and Ramama, 1966). Both of these factors result in arrivals at distant geophones having onsets on the seismic record less distinct than those of closer geophone traces.

Second, excessive shotpoint offsets may make the planting of a geophone at the reverse shotpoint impractical. An arrangement employed by the Bureau of Mineral Resources in Australia for many years uses the shot firing cable to transmit the signal from a geophone at the reverse shotpoint, when it is not required for shot firing (Hawkins, 1961, p. 811). However, for large shotpoint offsets, this system may be inconvenient.

Third, it is possible that, even with a reciprocal geophone in place, the first arrival is from another deeper refractor.

Finally, disturbed ground caused by previous shotpoints may result in unknown significant irreversible delays, as shown by Domzalski (1956, p. 145). Although geophones are usually planted away from earlier shotpoints, the region of the disturbed ground is sometimes larger than expected.

Reciprocal times for distant shots

With large shotpoint offsets, the recording of a reciprocal time is usually not convenient. Although reciprocal time errors of any magnitude can be corrected, it is still preferable to use as accurate a value as possible. Not only do graphical presentation of time depths and velocity analysis functions then lie within convenient sized regions on diagrams, but average velocity calculations can be made at early stages in the interpretation sequence.

A convenient approximation for the reciprocal time for two distant shots C and D (see Figure 32), when an intermediate value between A and B is known, is

$$t_{CD} = t_{CB} + t_{DA} - t_{AB}. \tag{33}$$

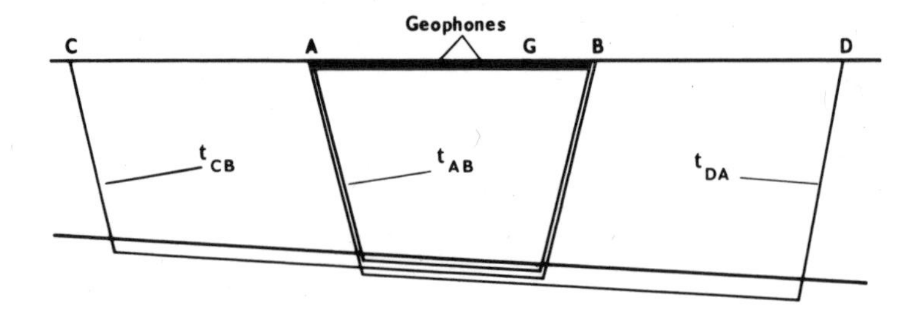

FIG. 32. Raypaths used in the calculation of a reciprocal time from an intermediate value.

Corrections for errors in reciprocal times

The error in the time-depth Δt_G resulting from an error in the reciprocal time Δt_{AB} is obtained by differentiation of equation (10), viz,

$$\Delta t_G = -\tfrac{1}{2}\,\Delta t_{AB}. \tag{34}$$

The time-depth at the point displaced $\frac{1}{2}\,XY$ from the shotpoint at A, toward the shotpoint at B, is given by the limit

$$[t_v]_{x=0} = 1/2\,[t_{AY} - t_{BX} + t_{AB}]_{x=0}. \tag{15}$$

This limit is the intercept on the time axis of the expression for the velocity analysis function t_v, and the error in the reciprocal time is independent of x. Thus we can differentiate under the limit to obtain

$$\Delta\,[t_v]_{x=0} = \tfrac{1}{2}\,\Delta\,t_{AB}. \tag{35}$$

Combining equations (34) and (35) we obtain

$$\Delta\,t_G = -\,\Delta\,[t_v]_{x=0}. \tag{36}$$

Thus the error in time-depths at all geophones Δt_G, caused by an error in the reciprocal time, has the same magnitude but opposite sign to the error in the time-depths at the shotpoints $\Delta\,[t_v]_{x=0}$ caused by the same error in the reciprocal time.

Adjustment in the manner described by equation (36) preserves the traveltime curves.

$$t_{AY} = t_G + [t_v]_{x=0} + AY/V_n' \tag{37}$$

$$= \text{corrected } t_G + \text{corrected } [t_v]_{x=0} + AY/V_n'. \tag{38}$$

An example of this method of adjustment is shown in Figure 33. Velocity analyses and time-depths are plotted for several values of the reciprocal time. To avoid confusion, the following convention has been used.

Time-depths below geophones are joined by solid lines. These lines can be considered to be rigid couplings which ensure that these time-depths are adjusted together in the same direction during interpretation.

Time-depths below shotpoints are joined to the time-depths below the geophones, formed from times from those shots, by a broken line. The broken

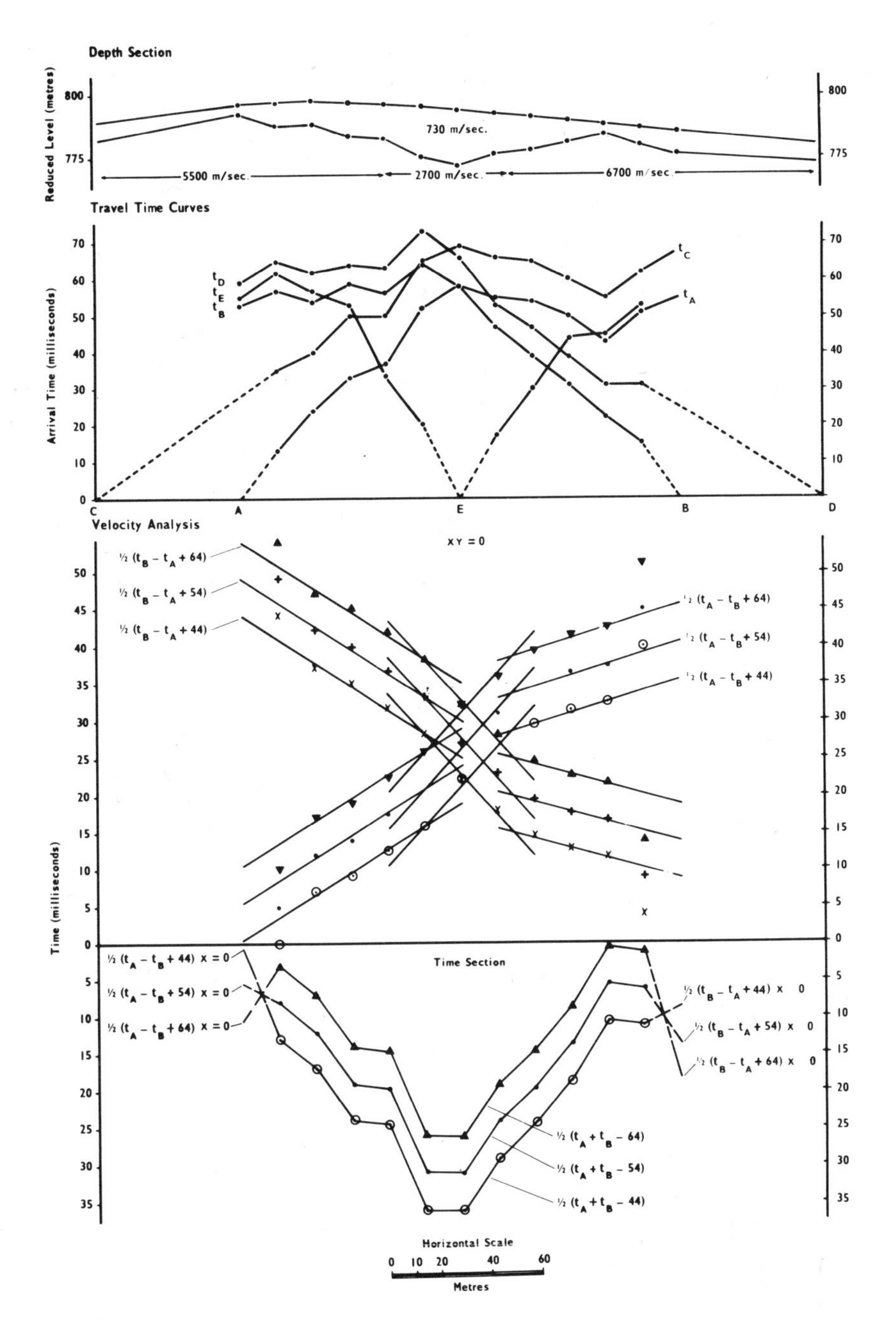

FIG. 33. Velocity analysis functions and time-depths, calculated from the travel-time curves from shots at A and B, are presented for three reciprocal times. Changes in the reciprocal time result in simple vertical displacements. The geophone time-depths are shifted in the opposite direction to the shotpoint time-depths, in the "butterfly adjustment" method.

line can be considered to be a flexible coupling and indicates that these time-depths are adjusted in the opposite direction to those below the geophones.

The intercept time

In theory, there is a wide range of values which the reciprocal time can assume, while still preserving the traveltimes. With adjacent or overlapping spreads this ambiguity can be resolved if a shotpoint location coincides with a geophone location. Adjustment, using equation (36), can then be made so that as close agreement as possible is obtained between geophone and shotpoint time-depths at the same locations.

However, with single geophone spreads, additional information is necessary to resolve the ambiguity. This additional information is the intercept time. Time-depths are adjusted so that as close agreement as possible with half-intercept times is obtained, while still using equation (36).

The conventional intercept time is defined as the intercept on the time axis, obtained by projecting the segment of the traveltime curve corresponding to that layer back to the shotpoint.

However, with undulating refractors or lateral velocity variations, the traveltime curves depart from sets of simple straight line segments, and as a result, the intercept times are obtained from lines of best fit. The accuracy of the intercept time determined by this method is therefore reduced, and this limits the value of adjusting time-depths to half-intercept times using equation (36).

However, it is possible to define a generalized intercept time that overcomes the above problem. Using the symbols of Figure 32, a generalized intercept time at A is expressed as

$$t_{CA} - (t_{CG} - t_{AG}). \tag{39}$$

Although the conventional intercept time has been defined as the intercept on the time axis of the traveltime curve at the shotpoint, it can be considered to be the arrival time at the shotpoint for a critically refracted ray. In the determination of the conventional intercept time, this fictitious arrival time is obtained by extrapolation, which is only valid for plane layers. However, the definition in equation (39) utilizes the well-known property that traveltime curves in layers with uniform velocities are parallel. The term $(t_{CG} - t_{AG})$ in equation (39) is the separation between the curves, and equation (39) effectively displaces the traveltime curve from the shotpoint at C to coincide with the traveltime curve for the shotpoint at A, to obtain the fictitious arrival time.

Not only does equation (39) provide a value identical to that obtained by extrapolation for the plane layer case, but it also provides a meaningful value when undulating refractors, or lateral velocity variations, exist such as are shown in Figure 33. The half-generalized intercept time at E is 30.5 msec. This compares with 31 msec for the time-depth at this point using a reciprocal time of 54 msec, which appears to be the correct value.

The similarity between equation (39) and equation (10) for zero XY emphasizes the similarity between the intercept time method and the conventional reciprocal method. It will be noted that while the formation of time-depths with equation (10) requires forward and reverse arrival times, the formation of generalized intercept times with equation (39) only requires traveltime data recorded in one direction.

Chapter 10
The time section

The construction of time sections using the GRM is a very convenient and powerful intermediate stage in the processing and interpretation of seismic refraction data.

The time section is an orthogonal plot of time-depths and half-intercept times below the points on the surface to which each refers. The horizontal axis represents the seismic line on the earth's surface. For a geophone time-depth [equation (10)], the reference point is the midpoint of XY; for a time-depth near a shotpoint [equation (15)], it is $1/2$ XY from that shotpoint toward the reverse shotpoint; for the half-intercept time [equation (39)], it is below the shotpoint. The vertical axis has the units of time positive downward.

Time sections have been used previously in seismic refraction processing for the adjustment of delay times (Wyrobek, 1956; Pakiser and Black, 1957; Layat, 1967). However, the most common use of time sections is in the seismic reflection method (Dobrin, 1976, p. 236).

The time section is an extremely convenient work area for seismic refraction processing because (1) it is not affected by uncertainties in determining velocities above refractors, (2) it can be constructed while still preserving consistency with the original traveltime data, and (3) it provides a criterion for field work requirements.

Uniqueness of the time section

For the calculation of a time-depth using equation (10), the only velocity required is that of the refractor. This can be readily obtained using equation (6). The velocities of layers above the refractor being mapped are not required to form equation (10).

However, to convert time-depths into depths using equations (12) and (14), the seismic velocities of all layers above that being mapped are required. Since hidden layers within blind zones, velocity inversions, nonuniform layers, and lateral velocity variations can occur, many different depth sections can produce the same time section.

In general, a unique time section can be constructed from the seismic data alone, provided there are sufficient data to resolve ambiguities, but additional data such as drillhole controls are necessary to derive a unique depth section.

Consistency between data and interpretation

Many methods used for the interpretation of seismic refraction data are based on approximations and assumptions, such as negligible dips, constant refractor velocities, etc. As a result, traveltimes calculated from the interpreted depth section need not necessarily agree with the observed traveltime data, i.e., the interpretation is not consistent with the data.

To achieve consistency, the interpretation can be adjusted until the calculated and observed traveltimes agree. Scott (1973) described a computer program in which a delay time method interpretation is adjusted, through a number of iterations, until the traveltimes of the interpretation determined by a ray-tracing routine agree with the observed traveltime data.

Other methods, such as wavefront constructions (Thornburgh, 1930; Rockwell, 1967) and graphical methods (Slotnick, 1950; Tarrant, 1956; Hales, 1958; Stulken, 1967) use the laws of geometrical optics, and are therefore mathematically exact. However, the accuracy of ray tracing, wavefront construction, and graphical methods depends upon the recognition and definition of all layers above the refractor being mapped. Unless all layers are defined, then these methods simply produce an interpretation which is mathematically consistent, but not necessarily physically relevant (see also MacPhail, 1967, p. 260).

As shown previously, the method of adjustment with equation (36) preserves the traveltime curves. However, it operates within the time domain and it does not require any knowledge of the overburden velocity distribution. It is therefore a more general and accurate method of obtaining consistency between data and interpretation than ray tracing, wavefront construction, or graphical methods.

Field data requirements

In general, the complexity of the time section is directly dependent upon that of the geologic section. If the geologic environment is reasonably uniform, then a large amount of interpolation in the time section is probably valid. However, if the geologic environment is complex, then in order to map all refractors, a thorough field program which will produce a time section with the desired level of interpolation commensurate with the complexity of the geology is required.

For the formation of time-depths and velocity analysis functions, forward and reverse segments of the traveltime curves over a common interval of the refractor are required. Therefore, shot and geophone arrangements which permit time-depths and velocity analysis calculations are required, and the mere production of complete traveltime curves is insufficient.

Chapter 11
An interpretation routine

One of the well-known shortcomings of the seismic refraction method is the disproportionate length of time required to produce final detailed results, when compared with the time taken to acquire data. Processing is not a major problem because digital computers and plotters facilitate rapid computation and presentation of results (Peraldi and Clement, 1972; Scott, 1973). Most delays occur in the interpretation phase, which requires high levels of expertise and judgment (Dobrin, 1976, p. 294).

In the majority of refraction interpretation routines (e.g., Hawkins, 1961, p. 811, 812), processing follows interpretation of data. This procedure is adequate when there are only a few spreads to interpret or when the processing facilities are convenient. However, when there are large amounts of data, or when consistency with cross-lines or drillholes is required, then reprocessing following each reinterpretation may be inconvenient as well as costly.

With the GRM, the interpretation routine involves examination of both the basic traveltime data and the processed data, viz., the velocity analysis functions and time-depths, in order to recognize optimum XY-values, surface irregularities, etc. Therefore, for the most efficient use of the GRM, the interpretation phase should *follow* the data processing phase. The interpretation phase then requires an editing phase, where meaningless computations are removed from further consideration.

Data processing

Processing using the GRM is best carried out using a computer and plotter because of the large number of points produced when a range of XY-values is used. A program has been published by Hatherly (1976) for the processing and plotting of seismic refraction data using the GRM.

The data processing phase includes the measurement and plotting of traveltimes, and the computing and plotting of the velocity analysis functions and time-depths. Traveltimes can be measured by hand, by crosscorrelation methods (Peraldi and Clement, 1972), or by statistical methods (Hatherly, 1979).

In general, it is prudent to make a rough plot of the traveltime curves, either by hand, on a computer line printer, or on a drum plotter. Any obvious errors in traveltimes, as well as up-hole and time-break errors, shown in reciprocal time mismatches, can then be corrected. Also, the rough plot permits a choice of shot pairs which may give useful time-depths and velocity analysis functions.

Usually two or three refractors can be mapped depending upon the separation of the shot pairs selected.

The traveltime data, and the shot pairs and corresponding reciprocal times are submitted to the computer, and the traveltime curves, velocity analysis functions, and time-depths are computed and drafted with a flat bed plotter on transparent film. For a given shot pair, different reciprocal times are used with each XY-value so that the velocity analysis functions and time-depths are separated for each XY-value in the plot. Several contact prints are made for the interpretation phase.

Interpretation

In the interpretation phase each arrival time on the traveltime curves is assigned to a refractor, the velocity analysis functions and time-depths are edited, an XY-value is selected, refractor velocities are measured, time-depths are adjusted, and velocities are assigned to each layer.

The assigning of each arrival time to a refractor is by far the most important aspect of refraction interpretation, particularly in shallow investigations. Single traveltime curves are ambiguous, because an increase in apparent velocity, with an increase in distance from the shotpoint, may be caused by either a change in the dip, and/or the velocity in the same refractor, or by the recording of a deeper, higher-velocity refractor. Reverse shooting does not resolve this ambiguity when the change in slope occurs at a similar geophone location.

The ambiguity can be resolved by shooting at least two shots from different locations at each end of the geophone spread. The point of change of slope in the traveltime curve will be displaced horizontally toward the more distant shotpoint, if a deeper, higher-velocity refractor occurs. However, with a change of dip or velocity in the same refractor, the traveltime curve is displaced vertically. The GRM itself differentiates between a dip and/or a velocity change in the refractor.

An extremely convenient method of recording this interpretation is to trace over the traveltime curves with colors, with different colors corresponding to different refractors.

Equation (2) for the calculation of the velocity analysis function, and equation (10) for the calculation of time-depths, are only valid for arrivals from the same refractor. The velocity analysis function and time-depths must therefore be edited. Any computations which use arrivals from different refractors are marked accordingly, and are not considered further since they are meaningless.

The next step is the selection of appropriate XY-values, as described in Chapter 6.

From the velocity analysis functions, velocity and time-depths data are recovered, using equations (6) and (15). The time-depths, obtained from the velocity analysis functions and the geophone time-depths using equation (10), are then adjusted to agree as closely as possible with half-intercept times [equation (39)], using equation (36).

If all refractors have not been mapped, then it may be necessary to use half-intercept times, and to interpolate between them. The accuracy of interpolation will depend upon the complexity of the geology.

Finally, velocities are assigned to each layer. If all refractors have been mapped, and have reverse coverage, then equation (6) yields velocities for each layer. However, this does not always occur. Therefore, either an average velocity from equation (27), velocities determined from the traveltime curves, velocities determined from laboratory tests on geologic samples, or empirical velocities based on drillhole or other geologic control can be used.

Depth section

Depths are calculated from the time-depths and velocities shown in the time section. In general, the approximation of equation (14) is adequate. However, two aspects should be considered to gauge the verisimilitude of the final depth section.

First, the XY-value derived from the final depth section should be similar to that observed in the traveltime, velocity analysis, or time-depth data.

Second, the migrated depth section should be a smooth surface (see "Detail of refractor definition," p. 29). Arcs which do not form a smooth envelope suggest either faulting (see synthetic examples), incorrect velocities, or near-surface irregularities being incorrectly projected into deeper layers.

Wavelength considerations

One other factor which should be considered is the relationship between layer thicknesses and wavelengths of the seismic pulse. The GRM, like most methods of refraction interpretation, follows the ray theory approach. However, these methods of geometrical optics are only valid for wavelengths smaller than the lengths of the raypaths and thicknesses of the layers present (Grant and West, 1965, p. 128–132).

In the refraction method, both the source and the receiver must be several wavelengths from the interface (Grant and West, 1965, p. 183, 184), and the thickness of the refractor must at least equal one wavelength, or anomalous velocities will be measured (Press et al, 1954; Lavergne, 1961; Levin and Ingram, 1962; Donato, 1965).

The results of Levin and Ingram (1962) also provide a guide to the effects of a thin surface layer. Their model has a surface layer which is 2¼ inches thick with a seismic velocity of 7600 ft/sec. The approximate pulse frequency of 100 KHz (see page 755) gives a wavelength of about 1 inch. Although the surface layer thickness is more than twice the wavelength, the maximum intercept time recorded was only 40.5 μsec [see Levin and Ingram (1962) p. 759]. This corresponds with a calculated thickness of only 1¾ inches, using a refractor velocity of about 17,600 ft/sec. A similar result can be derived from case II (p. 395) of the model of Press et al (1954).

Therefore, depth sections which show relatively thin layers may not be accurate representations of the subsurface seismic velocity distribution.

Chapter 12
A field example

Within this presentation, fully defined synthetic models have been used to illustrate various features of the GRM. The benefits of this approach are obvious. Very few field examples have sufficient drillholes or other forms of sampling to guarantee the same control. Furthermore, models can be designed so that one aspect is emphasized, while other factors are held constant or are removed entirely.

However, like all methods of seismic refraction interpretation, the GRM will ultimately be judged on its ability to accommodate complexities of the real geologic environment. The seismic refraction survey at the Welcome Reef dam site (Hatherly, 1977) provides data which permit an assessment of the GRM, because field methods were appropriate. These methods included:

(1) The use of geophone separations considerably smaller than the depth to the deepest refractor;

(2) the measurement of arrival times to an accuracy of about one percent, despite the problems of introducing sufficient energy into the ground; and

(3) the use of sufficient shotpoints to resolve the ambiguity problem for the majority of layers.

A listing of locations and first arrival times for the refraction survey of the Welcome Reef dam site is given in Appendix B.

Geologic setting

The proposed Welcome Reef damsite is on the Shoalhaven River, approximately 50 km east of Canberra in southeastern Australia.

Outcrop in the area consists of interbedded, steeply dipping, Ordovician metasediments, which are mainly quartzites and phyllites. Weathering, particularly in the phyllites, can be extensive. In drillholes, the top of the weathered bedrock surface can often only be distinguished from the unconsolidated sediments by color changes, or the presence of remanent structures. These rocks are covered by 70 m of Tertiary and Quaternary unconsolidated alluvium, consisting of mainly loose, coarse sands and gravels, with interbedded layers of clay. Depths to the water table can be more than 30 m.

The presence of a conjugate northeasterly fault system within the area has been defined from aerial photographs.

The seismic refraction survey was carried out to determine the thickness

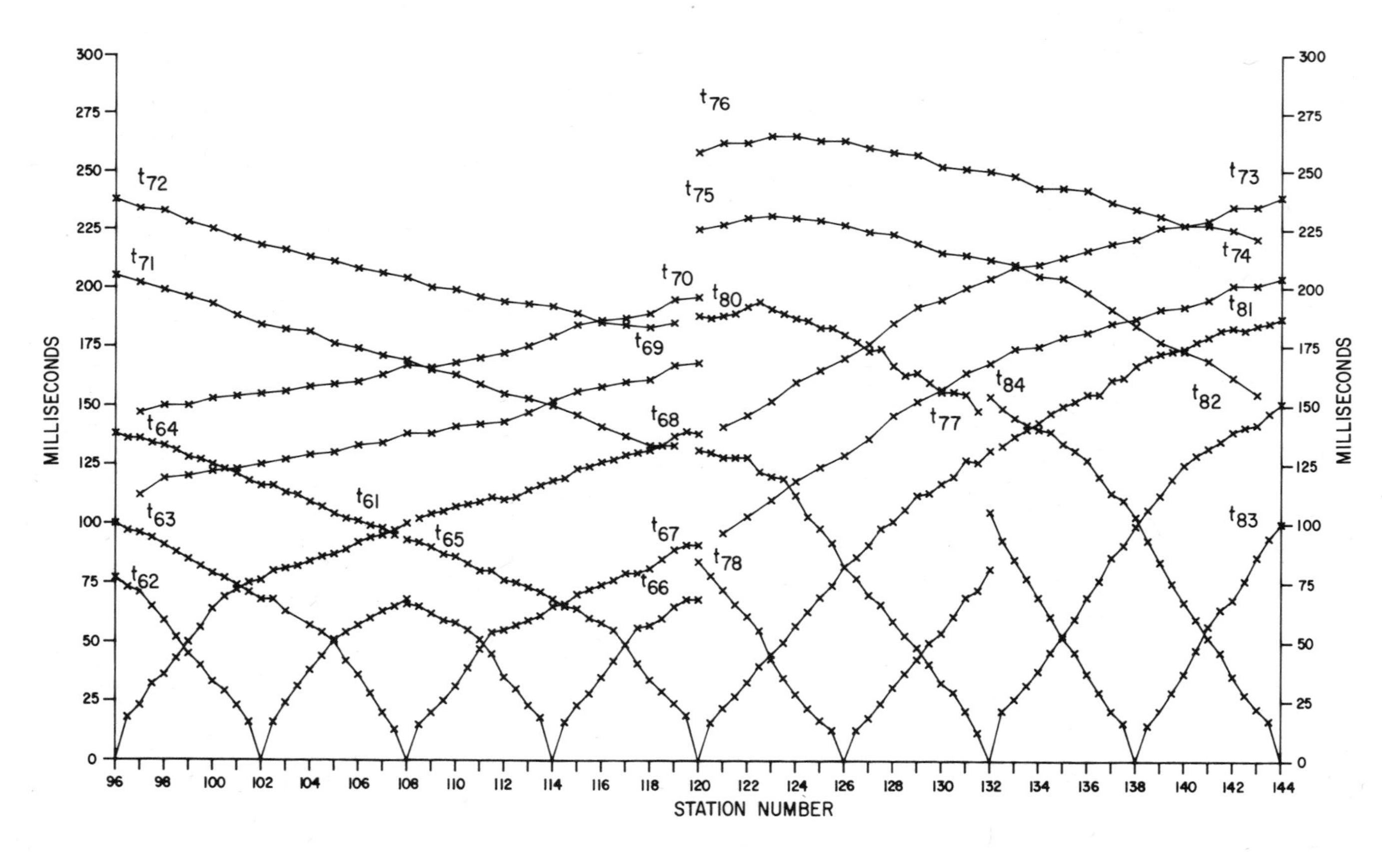

FIG. 34. Traveltime curves for part of line 3 where a fault is crossed. The station spacing is 10 m.

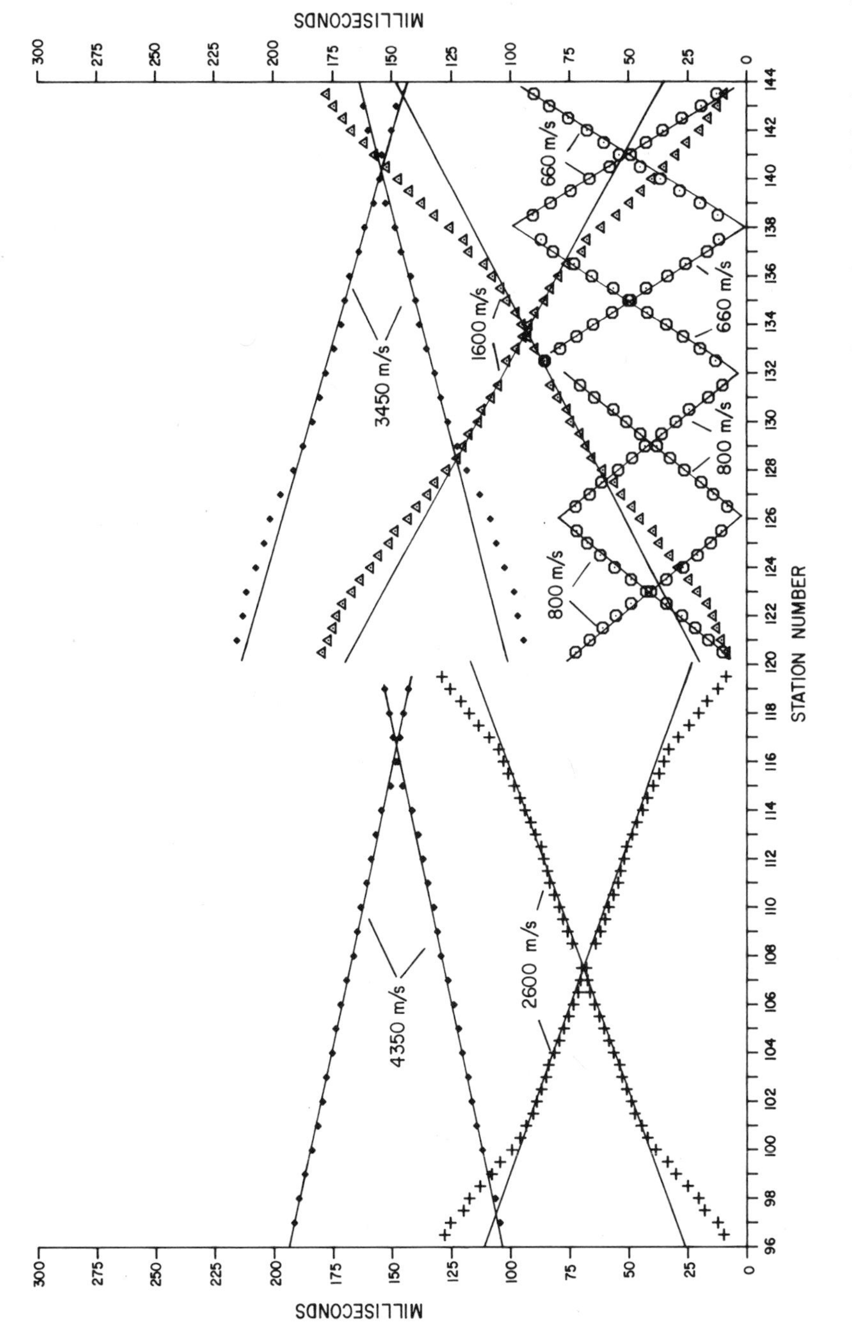

FIG. 35. Velocity analysis functions derived from data shown in Figure 34. The shot pairs to which the symbols refer are defined in Figure 36.

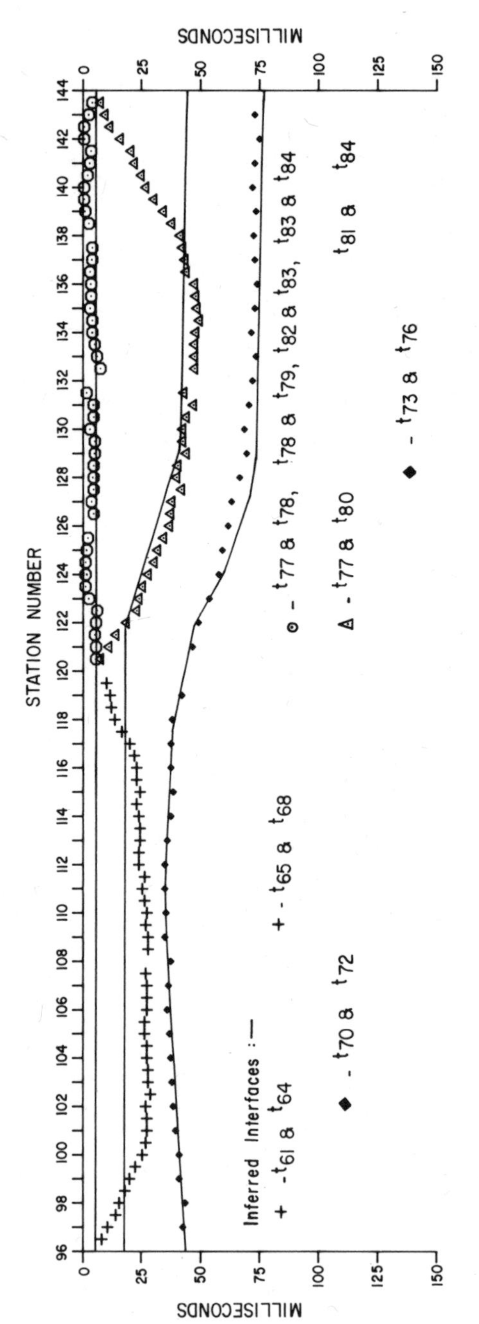

FIG. 36. Time section derived from data shown in Figure 34.

Table 6. Seismic model.

Layer number	Seismic velocity (m/sec)	Approximate thickness (m)	Geologic equivalent
1	340	1–2	Dry loose sand and soil
2	760–1000	10–30	Dry alluvium
3	1600–1850	15–65	Saturated unconsolidated sediments and completely weathered bedrock
4	2300–2700	20–50	Moderately weathered bedrock
5	2500–5000		Bedrock

of alluvium and the location of any faults, shears, or any other low-velocity zones in bedrock. All of these features could form leakage paths over a wide area around the left abutment.

Seismic velocity stratification

In Table 6, the velocity stratification obtained from previous seismic refraction surveys in the area is presented. Five layers have been recognized and can be correlated with the general features of the geology as described previously.

The previous seismic refraction results had shown that the third and fourth layers could occur as hidden layers when they were sufficiently thin.

Initial interpretation

An inspection of the traveltime curves in Figure 34 shows that the first, second, and fifth layers can be readily recognized. However, layers three and

Table 7. Initial interpretation.

Station 108			Station 144		
Seismic velocity (m/sec)	Time-depth or half-intercept time (sec)	Thickness (m)	Seismic velocity (m/sec)	Time-depth or half-intercept time (sec)	Thickness (m)
340		1.13	340		1.50
800	.003	11.79	800	.004	32.96
1600	.016	5.47	1600	.040	27.23
2600	.020	52.00	2600	.057	53.18
4350	.037		3450	.073	
Computed XY = 86.45 m			Computed XY = 166.47 m		

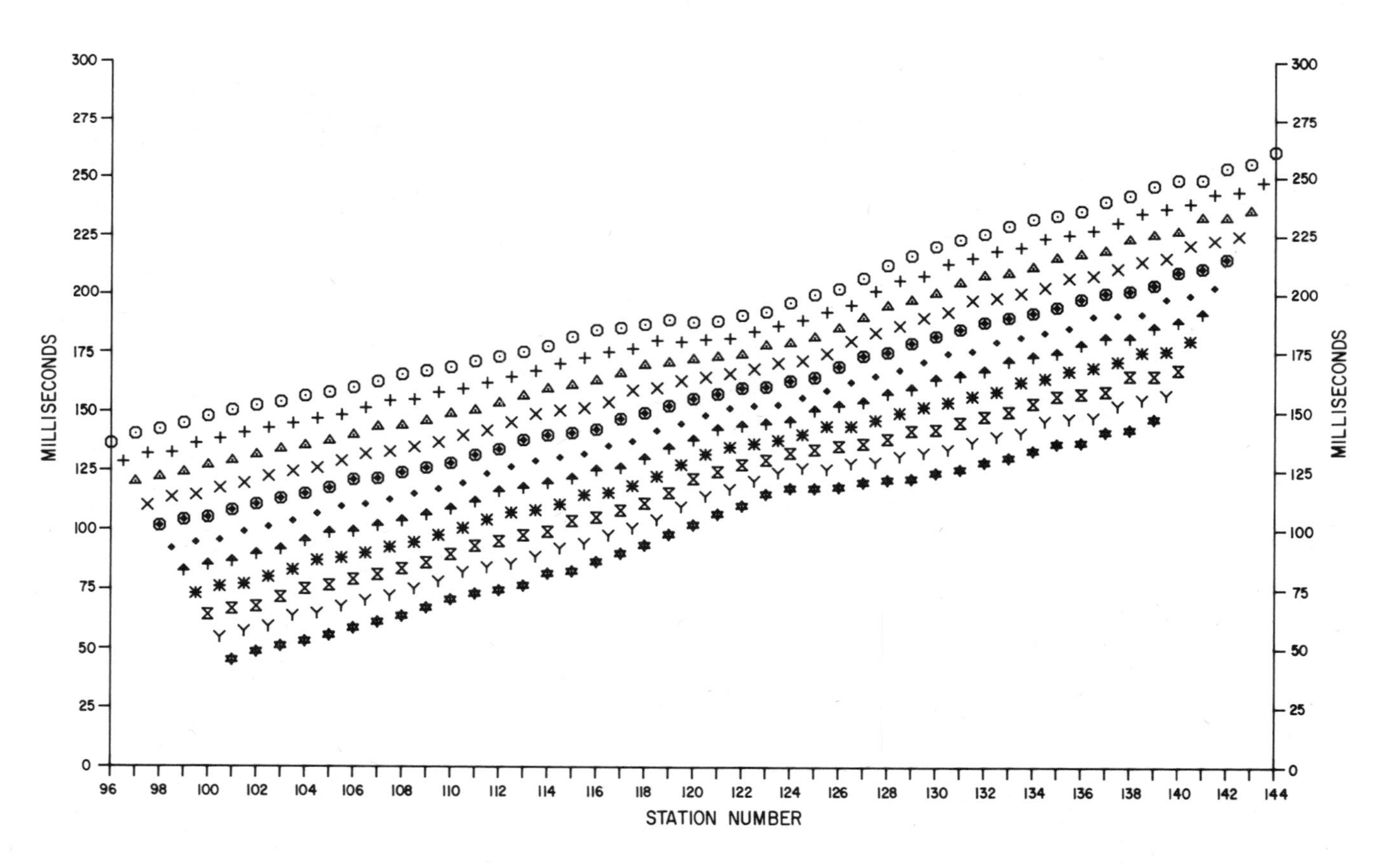

FIG. 37. Velocity analysis functions for XY-values from 0 to 100 m formed from traveltime curves obtained by combining shot numbers 70 and 73, and 72 and 76.

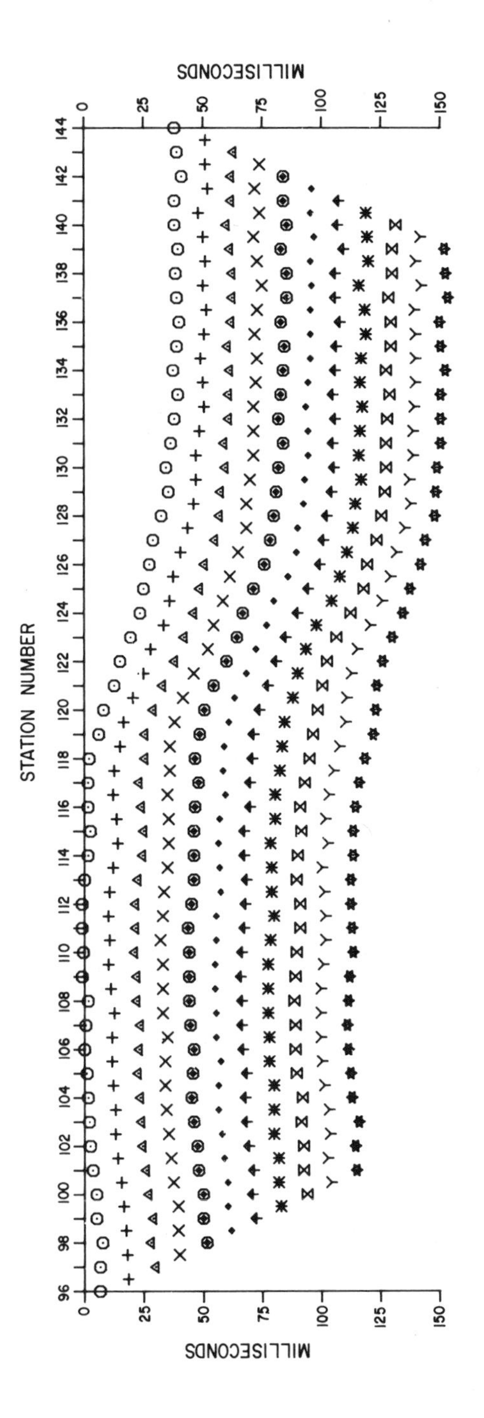

FIG. 38. Time-depths for XY-values from 0 to 100 m, formed from traveltime curves obtained by combining shot numbers 70 and 73, and 72 and 76.

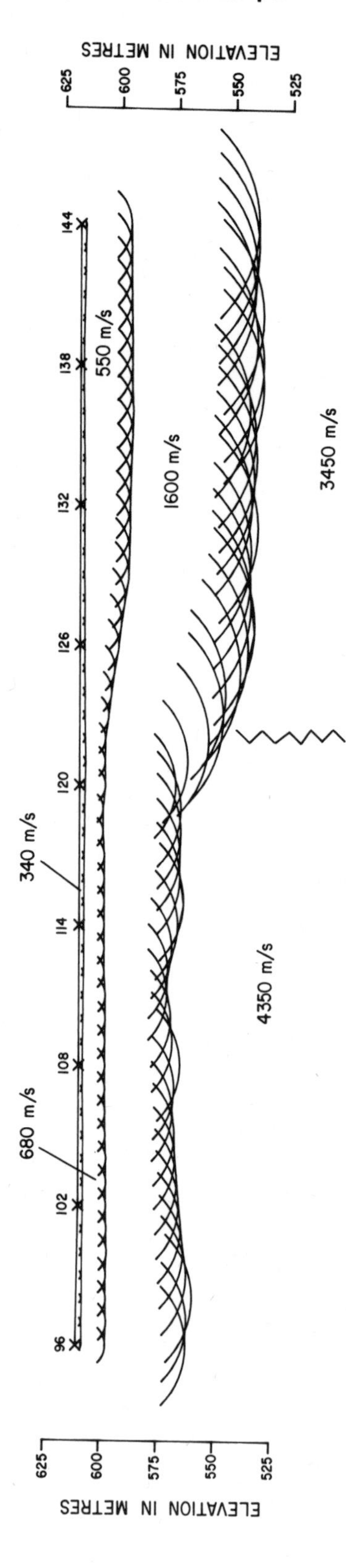

FIG. 39. Depth section derived from the data and processed data shown in Figures 34 to 38.

four, which are representative of saturated alluvium and weathered bedrock, cannot be easily separated.

Nevertheless, all five layers have been fitted to the traveltime curves, and the initial interpretation, together with the *XY*-values obtained with equation (20), are summarized in Table 7. Layer velocities have been taken from Figure 35, where a zero *XY* has been used throughout. The time-depth section derived from the data is shown in Figure 36.

Determination of *XY*-values

The traveltime data for shot numbers 70 and 73 have been adjusted or phantomed so that a single traveltime curve from one shotpoint is formed. It is used with a similar curve formed from shot numbers 72 and 76 to compute GRM parameters of *XY*-values ranging from zero (the upper set of points) to 100 m (the lower set of points) in Figures 37 and 38. The depth section is shown in Figure 39.

Optimum *XY*-values can be obtained in the following manner. The time-depths in Figure 38 show a rapid deepening of the refractor between stations 120 and 126. To the left of station 122.5 in Figure 37, the points for a 30-m *XY*-value show the best approximation to a straight line. This can be seen by either drawing lines through the points, or by covering the points between stations 123 and 144 with a sheet of paper and looking along the remaining points at grazing incidence.

In a similar manner, an *XY*-value of 60 m can be obtained for the points to the right of station 122.5. These values are less than half those computed from the preliminary depth section (see Table 7).

Revised interpretation

Drillholes located at stations 112 and 144 showed that depths to the water table were consistently shallower than those computed to the top of the third layer. This layer has a seismic velocity of 1600 m/sec, which is representative of saturated sediments.

A reassessment of the field records was made. Originally cycle skipping, which occurred on arrivals from the second layer, was interpreted as shingling associated with a sequence of interbedded sand and clays (Cassinis and Borgonovi, 1966). However, this phenomenon can also be explained by the occurrence of a velocity inversion (Greenhalgh, 1977). Traveltime curves for shot numbers 113 and 115, in Figure 40, are similar in appearance to those of Domzalski (1956, p. 153) for a reversal of velocity in the overburden. Accordingly, the seismic velocity of the second layer was adjusted until the calculated depths to the third layer were the same as those to the measured depth of the water table in the drillholes.

The next step in the revision of the interpretation was to determine an average velocity for the layers between the water table and bedrock. To do this, the contributions of the two layers above the water table were removed from the time-depths and the observed *XY*-values [see equations (30) and (31)]. These

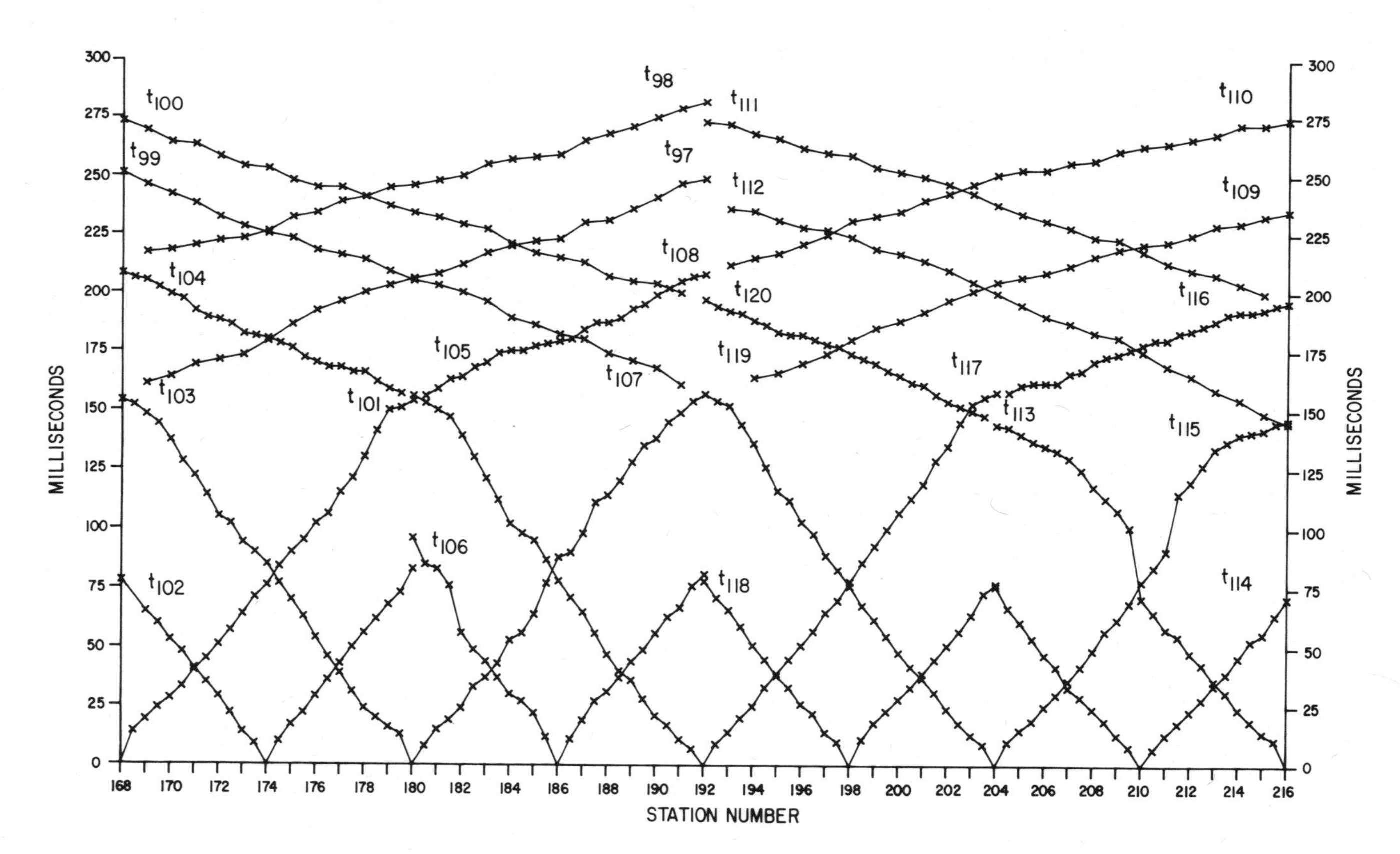

FIG. 40. Traveltime curves for another section of line 3. The station spacing is 10 m.

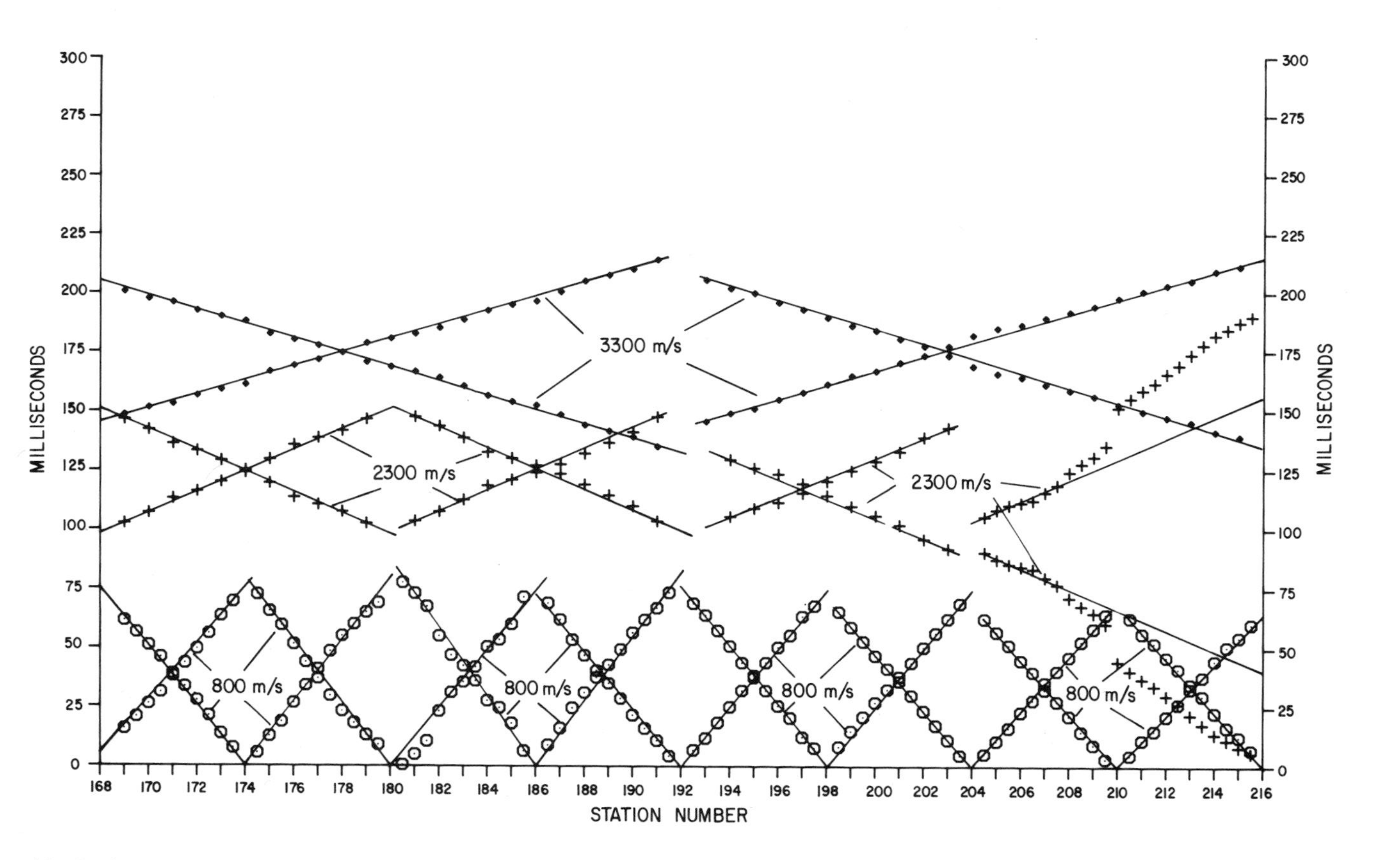

FIG. 41. Velocity analysis functions derived from data shown in Figure 40. The shot pairs to which the symbols refer are defined in Figure 42.

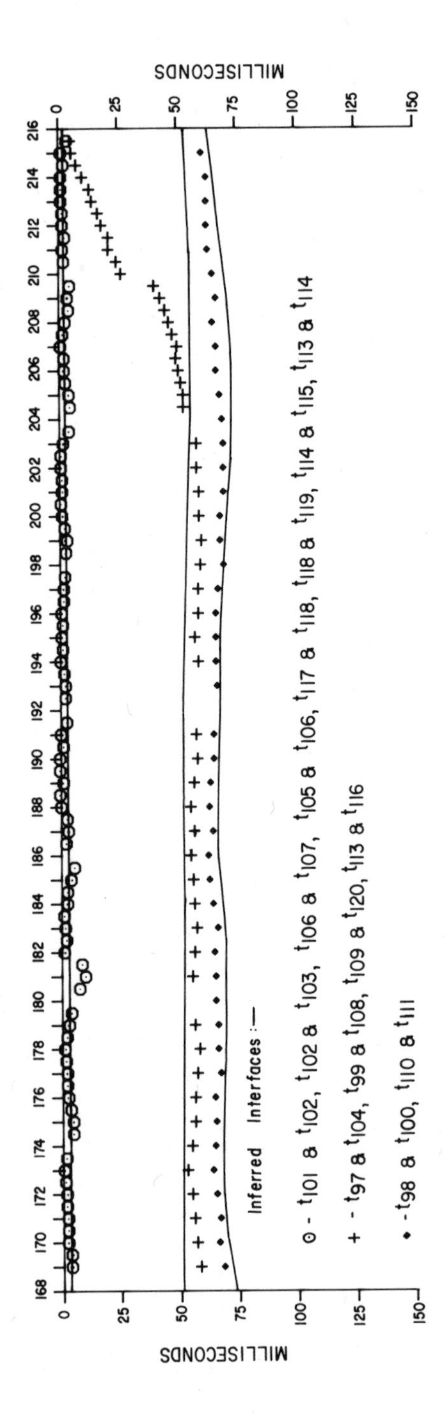

FIG. 42. Time section derived from data shown in Figure 40.

corrected values were used to obtain average velocities [equation (27)] and total thicknesses [equation (25)] for the layers between the water table and bedrock. The results are summarized in Table 8. A revised depth section plotted in Figure 39 is derived from the time section in Figure 36. The depth section shows a faulted bedrock, with an associated large change in thickness of overburden on the two sides of the fault.

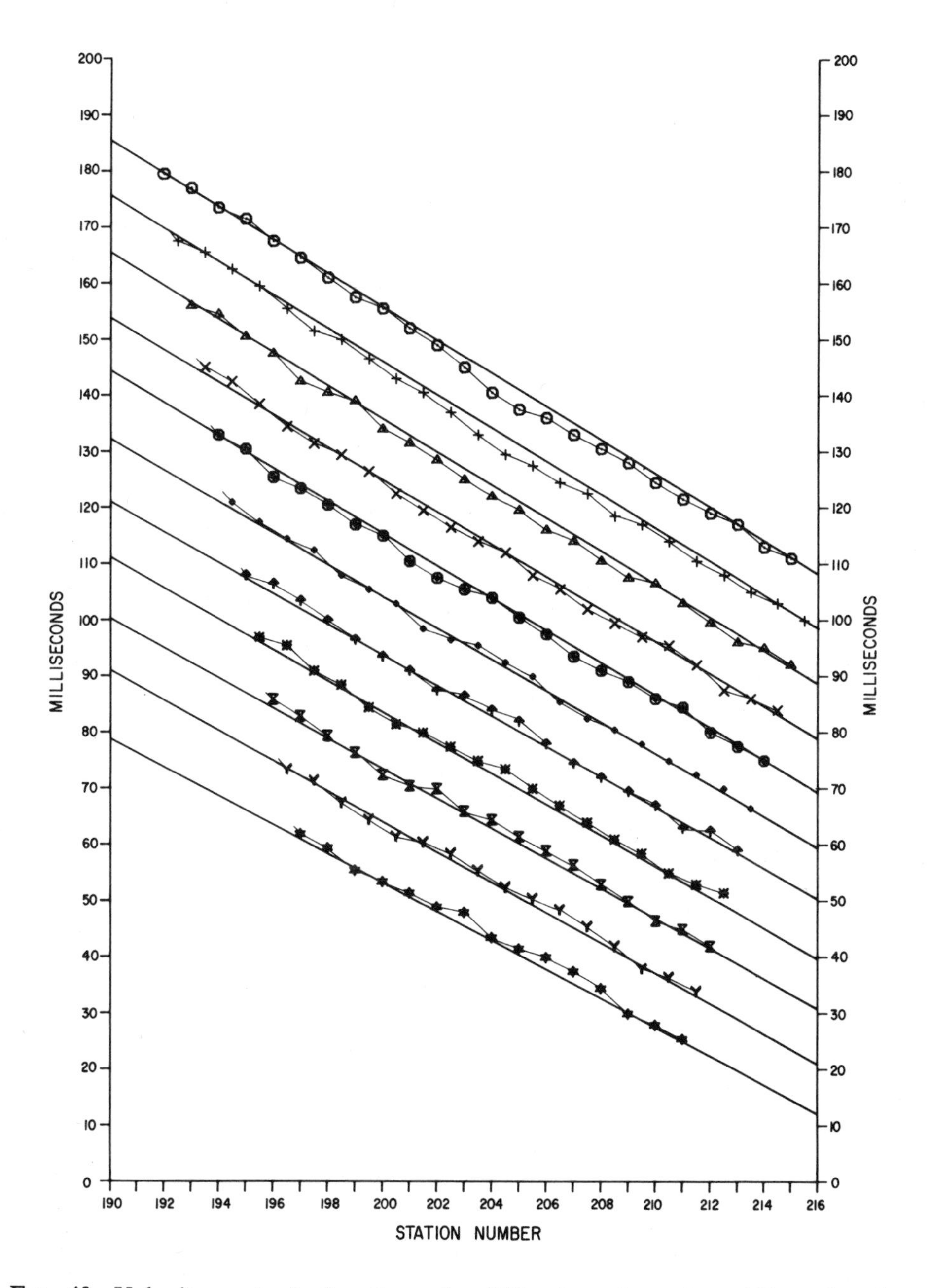

FIG. 43. Velocity analysis functions for XY-values from 0 to 100 m for shot numbers 110 and 111.

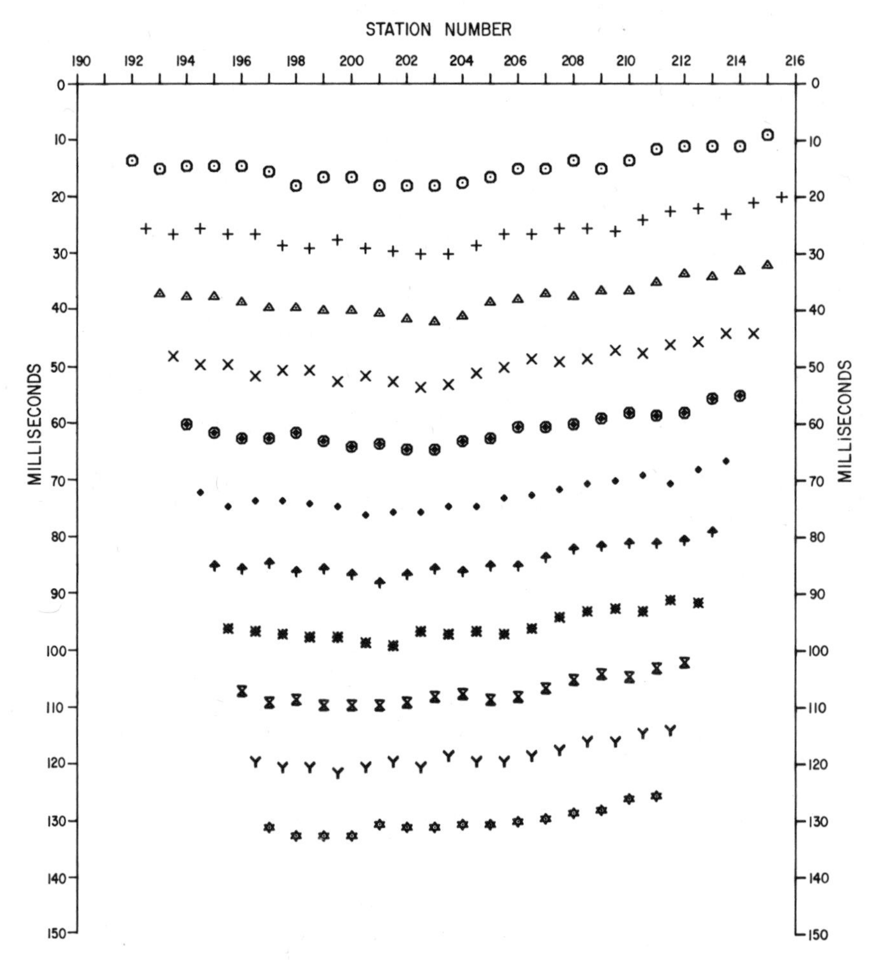

FIG. 44. Time-depths for XY-values from 0 to 100 m for shot numbers 110 and 111.

Table 8. Revised interpretation.

Station 108			Station 144		
Layer number	Seismic velocity (m/sec)	Thickness (m)	Layer number	Seismic velocity (m/sec)	Thickness (m)
1	340	1.18	1	340	1.73
2	680	9.48	2	550	20.52
XY correction: 3.18			XY correction: 6.97		
Time-depth correction: 0.017			Time-depth correction: 0.042		
Layer number	Seismic velocity (m/sec)	Thickness (m)	Layer number	Seismic velocity (m/sec)	Thickness (m)
3 and 4	1590	34.16	3 and 4	1535	53.35
5	4350		5	3450	

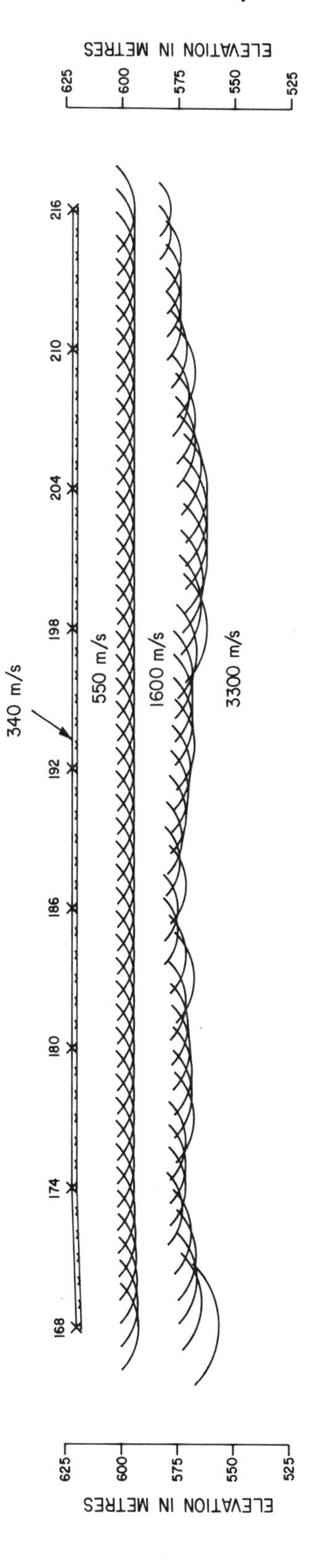

FIG. 45. Depth section derived from the data and processed data shown in Figures 40 to 44.

Allowing for errors of about ± 5 m in estimating XY, these results indicate that the fourth layer probably does not occur and that the change from completely weathered bedrock to reasonably fresh bedrock occurs over a fairly small interval.

Small refractor irregularities

The large change in depth to the refractor shown in the example in Figure 39 would rarely be encountered in most seismic refraction surveys. Nevertheless, small-scale irregularities can still provide optimum XY-values provided velocity analysis and time-depth data are clearly presented at suitable scales.

In Figures 40 to 42, traveltime, velocity analysis, and time-depth data are presented for two spreads recorded along the same line as the data shown in Figure 34. The processed data for shots 110 and 111 are presented in Figures 43 and 44. The time section in Figure 44 does not indicate any particular XY-value. However, in the velocity analysis in Figure 43, an XY-value of about 65 m can be recovered. The confidence in selecting this XY-value depends upon the recognition of the pattern shown in the velocity analysis data in Figure 26. In general, accurate arrival times are desirable. However, a more important factor is the geophone spacing. If the geophones are close together, then there will be more points defining the pattern of Figure 26. This data redundancy partially overcomes the need for accurate arrival times. The depth section in Figure 45 shows little relief in the refractor.

Accuracy of depth estimates

In the initial interpretation, each layer starting at the surface was defined sequentially. This approach resulted in depths which were more than 35 percent greater than those of the revised interpretation.

However, if an average velocity obtained from only observed XY, time-depth, and refractor velocity values had been used, then the depths would have been less than 16 percent greater than those of the revised interpretation. Even though the large thickness of low-velocity material above the water table reduces the accuracy of the average velocity method, an improvement in depth calculations still results when compared with the conventional interpretation approach.

A further improvement in depths could be obtained using the average velocity method, time-depth, and XY corrections as expressed in equations (30) and (31), and the seismic velocities and thicknesses in the initial interpretation (Table 7). The total depths would still have been less than 9 percent greater than those of the revised interpretation. However, the average velocity of the third and fourth layers would be 10 percent less, and accurate inferences on the existence or absence of hidden layers would not be possible.

Refractor velocity variations

In Figures 46 to 48, data and processed data are presented for part of a line several kilometers away from the previous examples.

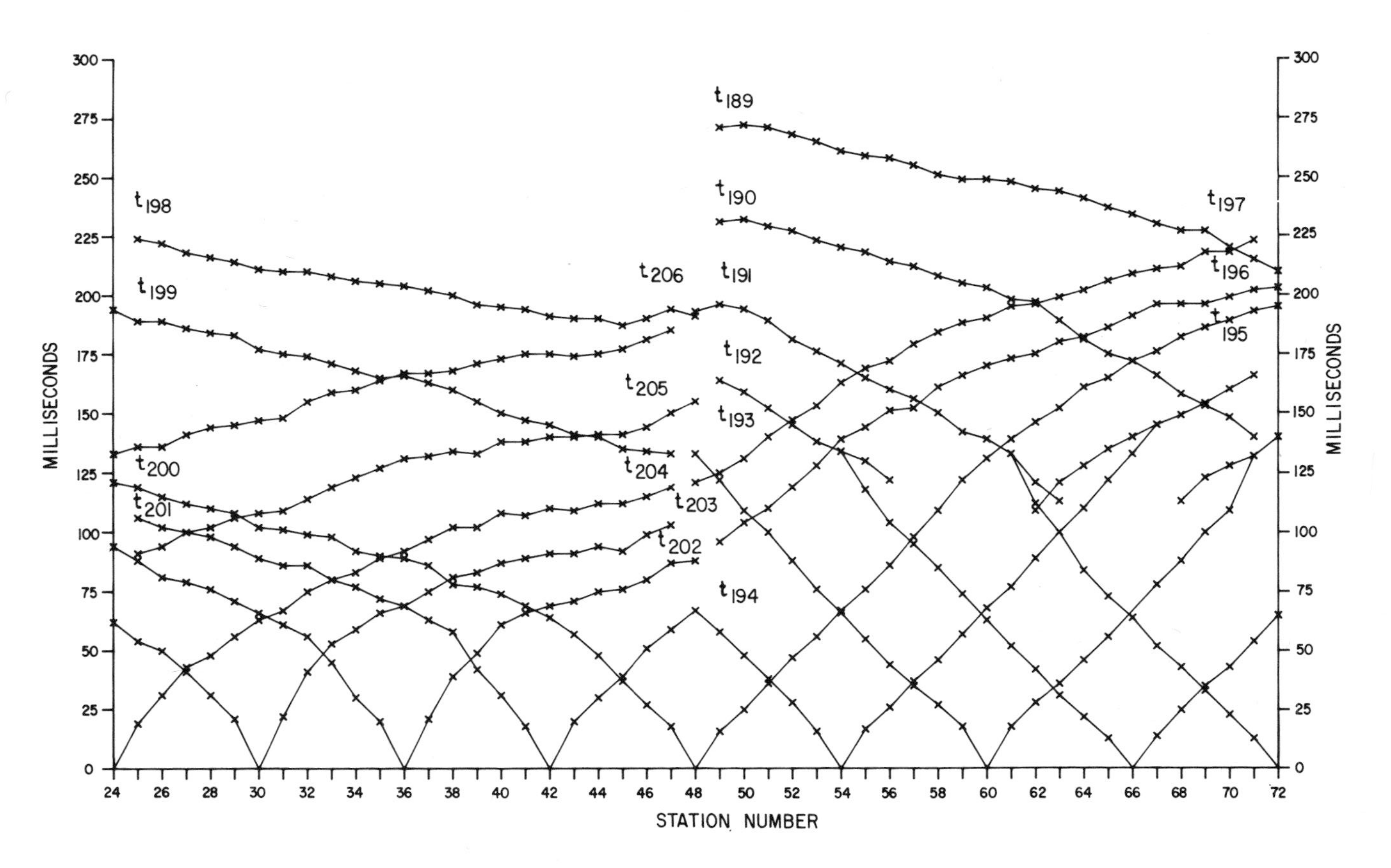

FIG. 46. Traveltime curves for part of line 4 where a fault is crossed. The station spacing is 10 m.

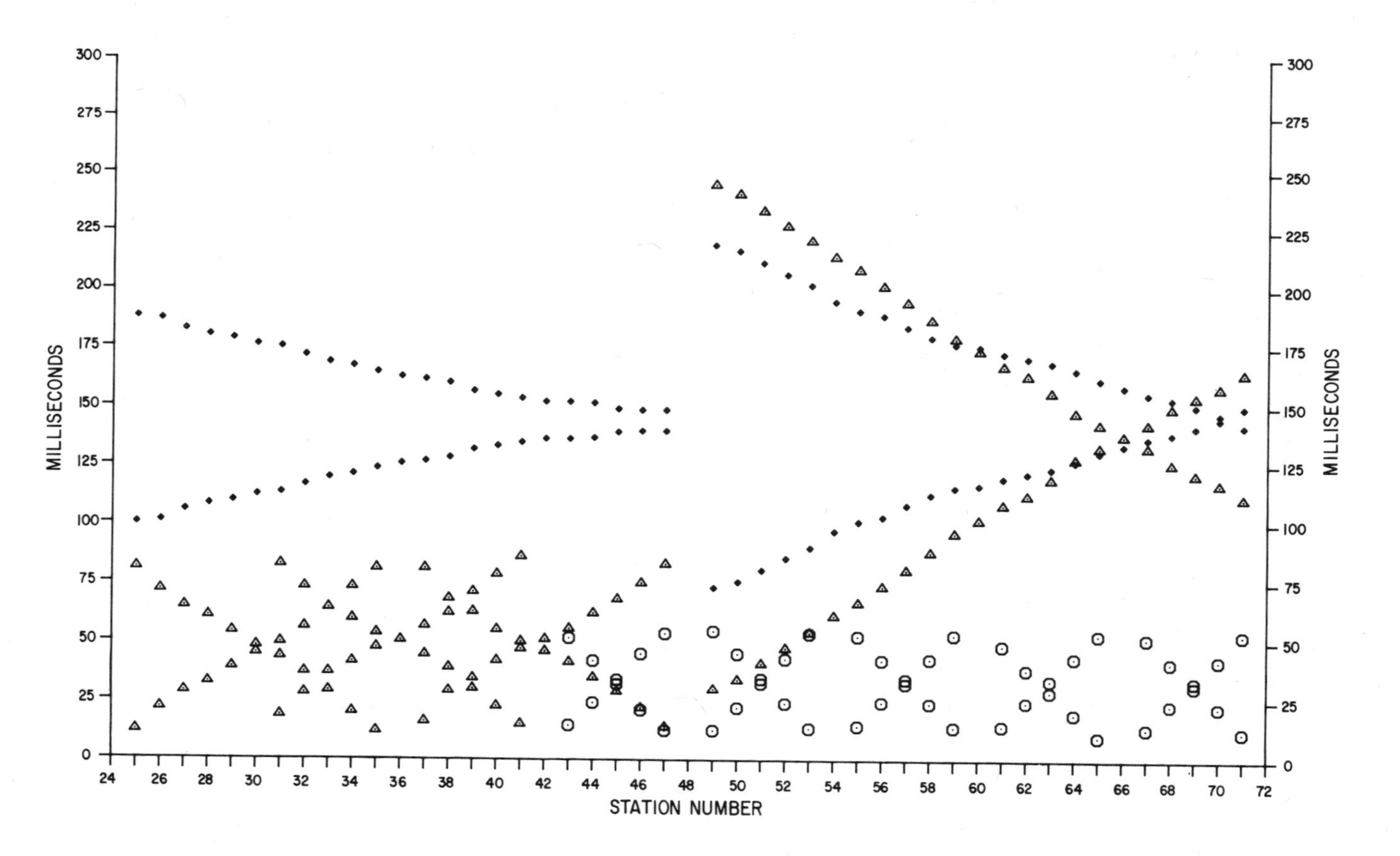

FIG. 47. Velocity analysis functions derived from the data shown in Figure 46. The shot pairs to which the symbols refer are defined in Figure 48. The reader is invited to make an interpretation.

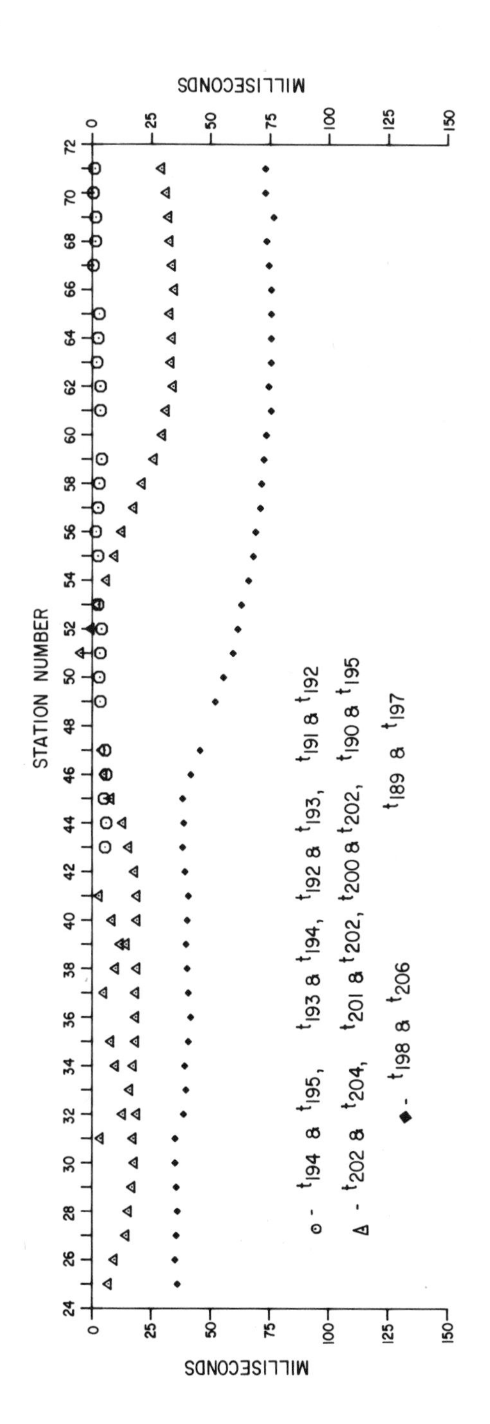

FIG. 48. Time section derived from data shown in Figure 46. The reader is invited to make an interpretation.

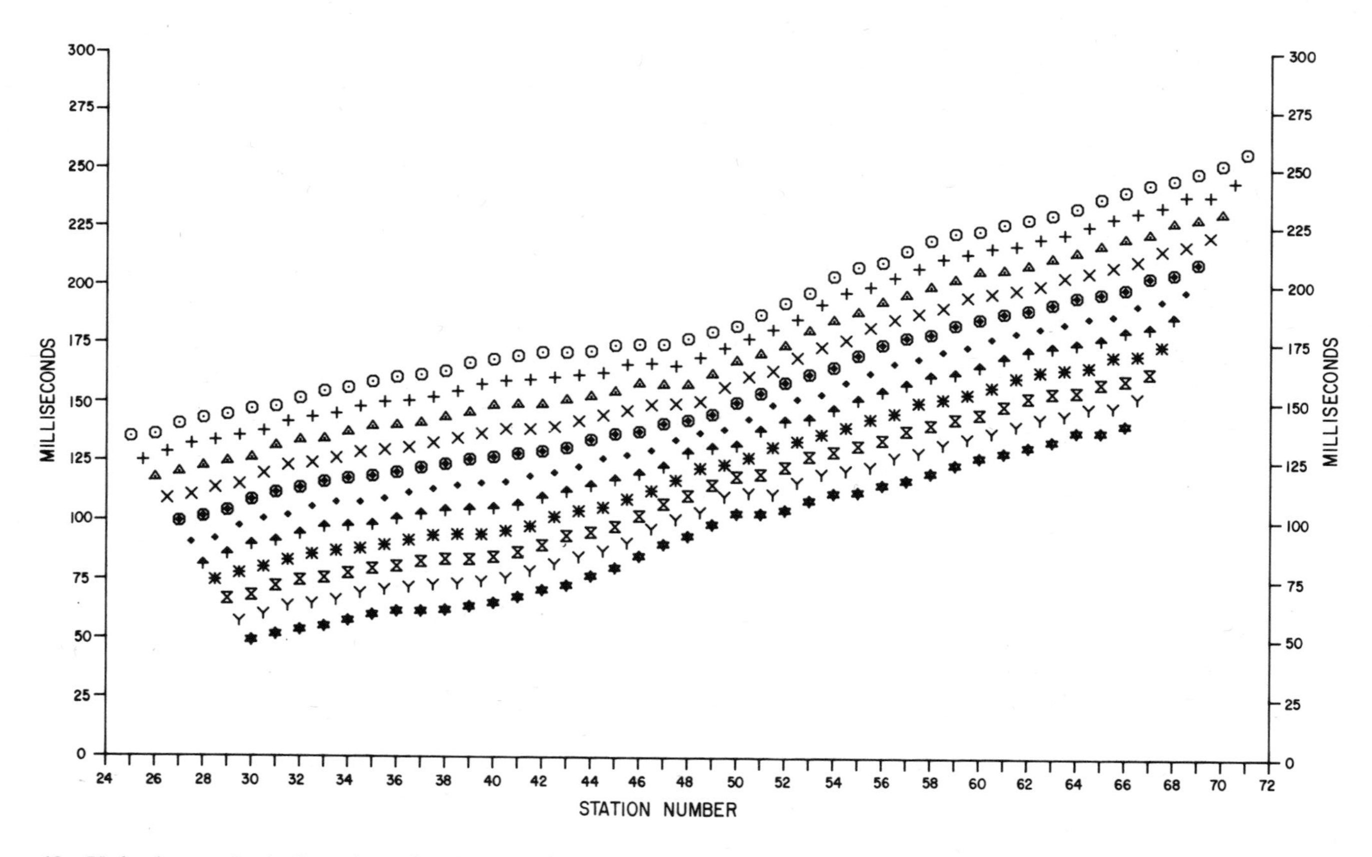

FIG. 49. Velocity analysis functions for XY-values from 0 to 100 m formed from traveltime curves obtained by combining shot numbers 189 and 197, and 198 and 206.

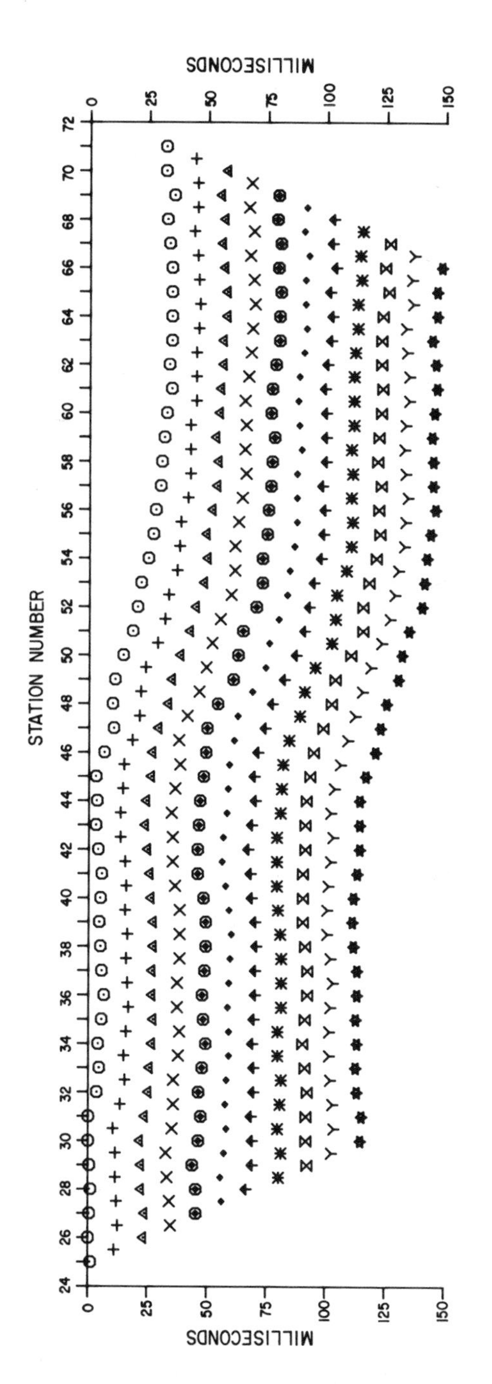

FIG. 50. Time-depths for XY-values from 0 to 100 m formed from traveltime curves obtained by combining shot numbers 189 and 197, and 198 and 206.

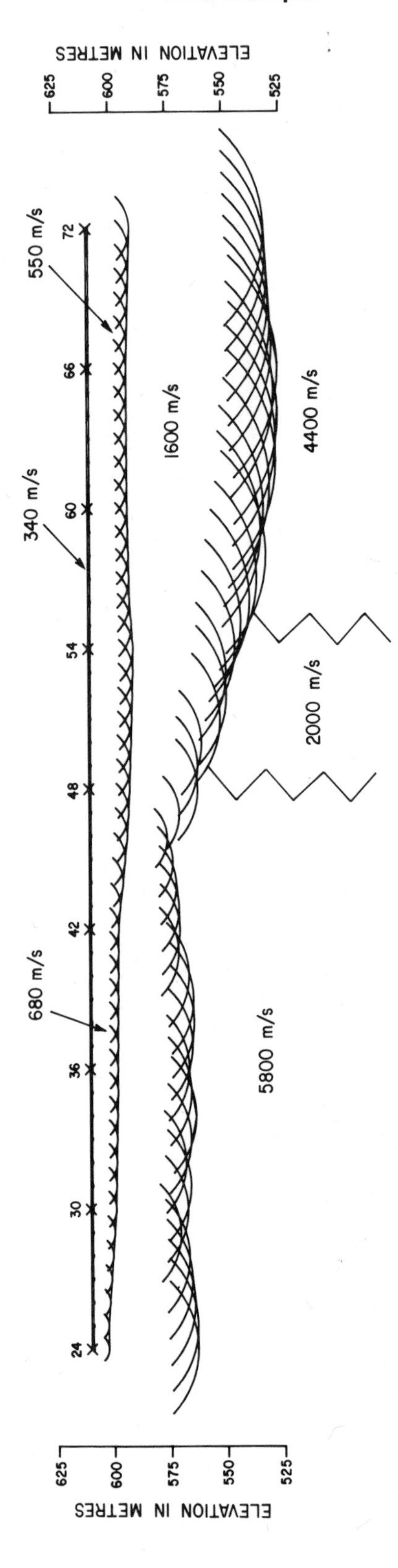

FIG. 51. Depth section derived from the data and processed data shown in Figures 46 to 50.

The velocity analysis data in Figure 49 (formed from shot numbers 189, 197, 198, and 206) show three refractor velocities. To the left of station 49, a velocity of 5800 m/sec can be obtained; to the right of station 55, the velocity is 4400 m/sec. A velocity of about 2000 m/sec is inferred for the 60-m interval between these stations. Figure 50 shows the time-depths for *XY*-values ranging between 0 and 100 m. In the interpreted depth section shown in Figure 51, the low-velocity interval may represent a zone of crushed and fragmented rock associated with a fault zone or a weathered phyllite band within quartzite. Irrespective of which explanation finally proves to be the most appropriate, this low velocity should be an important target in any subsequent drilling program.

However, definition of the extent of the low-velocity zone is difficult, and it rests largely on the ability to recognize the feature of Figure 26 when only a few points are available.

Summary

The Welcome Reef example provides an excellent examination of the GRM and the interpretation approach described in the previous chapter. The discussion has concentrated on the performance of the GRM in accommodating undetected layers and changing refractor depths and seismic velocities because these aspects can be verified reasonably objectively.

However, assessment of the efficacy of the interpretation approach is a little more subjective, and so it rests with each reader to form an opinion. The approach is expressed in the presentation of all data and processed data. If the reader considers there is sufficient information to assess adequately the GRM, then the value of the approach has been verified. This facility for critical evaluation is essential for the development of scientific thought, as well as the monitoring of production quality in routine investigations.

Chapter 13
The future

The GRM provides an integrated approach to seismic refraction interpretation cognizant of the realities of the geologic environment. These realities include undetected layers, and layers with variable thicknesses and seismic velocities.

For the most effective use of the GRM, accurate optimum XY-values are necessary. Improved accuracy may result through the application of statistical methods to parameters derived from first arrival times.

However, the powerful methods of time series analysis, such as those used in seismic reflection processing, probably hold more promise. In fact, even though many methods of seismic refraction interpretation are either special cases of, or closely resemble the GRM, it may be more prudent to pursue the similarities between the GRM and the seismic reflection method. These similarities include the correspondence between the time-depth and the one-way reflection time, as well as that between the average velocity and the RMS velocity. Furthermore, in each case it is possible to construct a time section which is independent of the accuracy of the seismic velocity determinations in the overlying layers.

Research and development into routine seismic refraction interpretation have been rather dormant for some time. I believe the development of time series analysis in seismic refraction processing to be long overdue.

Acknowledgments

I am indebted to Manus Foster and Kenneth Burke. Without their support and efforts, this work would not have been possible.

I am also indebted to my able colleague Peter Hatherly for carrying out the field work, processing of the example, and providing searching comments.

Comments by John Ringis, Adelmo Agostini, and Ted Tyne were invaluable.

The Cartographic Section of the Geological Survey of New South Wales prepared the figures, and Cheryl Smith prepared the manuscript more times than she would care to remember. Their contribution and patience are sincerely appreciated.

The work is published with the permission of the Under Secretary, New South Wales Department of Mineral Resources.

References

Acheson, C. H., 1963, Time-depth and velocity-depth relations in western Canada: Geophysics, v. 28, p. 894-909.

Adachi, R., 1954, On a proof of fundamental formula concerning refraction method of geophysical prospecting and some remarks: Kumamoto J. Sci., v. 2, p. 18–23.

Attewell, P. B., and Ramana, Y. V., 1966, Wave attenuation and interval friction as functions of frequency in rocks: Geophysics, v. 31, p. 1049–1056.

Barry, K. M., 1967, Delay time and its application to refraction profile interpretation, *in* Seismic refraction prospecting: A. W. Musgrave, Ed., SEG, Tulsa, p. 348–361.

Barthelmes, A. J., 1946, Application of continuous profiling to refraction shooting: Geophysics, v. 11, p. 24–42.

Bernabini, M., 1965, Alcune considerazioni sui rilievi sismici a piccole profondita: Bull. di Geofis. Teorica ed Applicata, v. 7, p. 106–118.

Berry, J. E., 1959, Acoustic velocity in porous media: Petroleum Trans. AIME, v. 216, p. 262–270.

Berry, M. J., 1971, Depth uncertainties from seismic first-arrival refraction studies: J. Geophys. Res., v. 76, p. 6464–6468.

Brandt, H., 1955, A study of the speed of sound in porous granular media: J. Appl. Mech., v. 22, p. 479–486.

Cassinis, R., and Borgonovi, L., 1966, Significance and implications at shingling in refraction records: Geophys. Prosp., v. 14, p. 547.

Chan, S. H., 1968, Nomograms for solving equations in multilayer and dipping layer cases: Geophys. Prosp., v. 16, p. 127–143.

Dix, V. H., 1956, Seismic prospecting for oil: New York, Harper and Row.

Dobrin, M. B., 1976, Introduction to geophysical prospecting, 3rd ed.: New York, McGraw-Hill Book Co., Inc.

Dooley, J. C., 1952, Calculation of depth and dip of several layers by seismic refraction methods: Austral. Bur. Min. Res. Geol. and Geophys., Bull. 19, Appendix, 9 p.

Domzalski, W., 1956, Some problems of shallow refraction investigations: Geophys. Prosp., v. 4, p. 140–166.

Donato, R. J., 1965, Measurements on the arrival refracted from a thin high-speed layer: Geophys. Prosp., v. 13, p. 387–404.

Duguid, J. O., 1968, Refraction determination of water table depth and alluvium thickness: Geophysics, v. 33, p. 481–488.

Duska, L., 1963, A rapid curved-path method for weathering and drift corrections: Geophysics, v. 28, p. 925–947.

Evjen, H. M., 1967, Outline of a system of refraction interpretation for monotonic increases of velocity with depth, *in* Seismic refraction prospecting: A. W. Musgrave, Ed., SEG. Tulsa, p. 290–294.

Ewing, M., Woollard, G. P., and Vine, A. C., 1939, Geophysical investigations in the emerged and submerged Atlantic Coastal Plain, Part 3, Barnegat Bay, New Jersey section: GSA Bull., v. 50, p. 257–296.

Faust, L. Y., 1951, Seismic velocity as a function of depth and geologic time: Geophysics, v. 16, p. 192–206.

———1953, A velocity function including lithologic variation: Geophysics, v. 18, p. 271–288.

Gardner, L. W., 1939, An areal plan of mapping subsurface structure by refraction shooting: Geophysics, v. 4, p. 247–259.

———1967, Refraction seismograph profile interpretation, *in* Seismic refraction prospecting: A. W. Musgrave, Ed., SEG, Tulsa, p. 338–347.

Gassmann, F., 1951, Elastic waves through a packing of spheres: Geophysics, v. 16, p. 673–685.

———1953, Note on "Elastic waves through a packing of spheres": Geophysics, v. 18, p. 269.

Goguel, F. M., 1951, Seismic refraction with variable velocity: Geophysics, v. 16, p. 81–101.

Grant, F. S., and West, G. F., 1965, Interpretation theory in applied geophysics: New York, McGraw-Hill Book Co., Inc.

Green, R., 1962, The hidden layer problem: Geophys. Prosp., v. 10, p. 166–170.

Green, R., and Steinhart, J. S., 1962, On crustal structure deduced from seismic time-distance curves: New Zealand J. Geol and Geophys., v. 5, p. 579–591.

———1974, The seismic refraction method—A review: Geoexpl., v. 12, p. 259–284.

Greenhalgh, S. A., 1977, Comments on the hidden layer problem in seismic refraction work: Geophys. Prosp., v. 25, p. 179–181.

Hagedoorn, J. G., 1955, Templates for fitting smooth velocity functions to seismic refraction and reflection data: Geophys. Prosp., v. 3, p. 325–338.

———1959, The plus-minus method of interpreting seismic refraction sections: Geophys. Prosp., v. 7, p. 158–182.

Hagiwara, T., and Omote, S., 1939, Land creep at Mt Tyausu-Yama (Determination of slip plane by seismic prospecting): Tokyo Univ. Earthquake Res. Inst. Bull., v. 17, p. 118–137.

Hales, F. W., 1958, An accurate graphical method for interpreting seismic refraction lines: Geophys. Prosp., v. 6, p. 285–294.

Hatherly, P. J., 1976, A Fortran IV programme for the reduction and plotting of seismic refraction data using the generalised reciprocal method: Rep. Geol. Surv. N.S.W., GS1976/236.

———1977, Seismic refraction investigations at the Welcome Reef damsite near Braidwood, N.S.W.: Rep. Geol Surv. N.S.W., GS1977/002.

———1979, Computer processing of seismic refraction data: Bull. Austral. SEG, v. 10, p. 217–218.

Hawkins, L. V., The reciprocal method of routine shallow seismic refraction investigations: Geophysics, v. 26, p. 806–819.

Hawkins, L. V., and Maggs, D., 1961, Nomograms for determining maximum errors and limiting conditions in seismic refraction surveys with blind zone problems: Geophys. Prosp., v. 9, p. 526–532.

———1962, Discussion on the problem of the hidden layer within the blind zone: Geophys. Prosp., v. 10, p. 548.

Heiland, C. A., 1963, Geophysical exploration: New York, Prentice-Hall, Inc.

Hollister, J. C., 1967, A curved path refractor method, *in* Seismic refraction prospecting: A. W. Musgrave, Ed., SEG, Tulsa, p. 217–230.

Iida, K., 1939, Velocity of elastic waves in granular substances: Tokyo Univ. Earthquake Res. Inst. Bull., v. 17, p. 783–807.

Jakosky, J. J., 1950, Exploration geophysics: Trija, Los Angeles.

Kaila, K. L., and Narain, H., 1970, Interpretation of seismic refraction data and the solution of the hidden layer problem: Geophysics, v. 35, p. 613–623.

Kaufman, H., 1953, Velocity functions in seismic prospecting: Geophysics, v. 18, p. 289–297.

Knox, W. A., 1967, Multilayer near-surface refraction computations, *in* Seismic refraction prospecting: A. W. Musgrave, Ed., SEG, Tulsa, p. 197–216.

Laski, J. D., 1973, Computation of the time-distance curve for a dipping refractor and velocity increasing with depth in the overburden: Geophys. Prosp., v. 21, p. 366–378.

———1978, Transformation of time-distance curves of diving waves into a multilayered model: Geophys. Prosp., v. 26, p. 48–81.

Lavergne, M., 1961, Etude sur modele ultransonique du probleme des couches minces sismique refraction: Geophys. Prosp., v. 9, p. 60–73.

Layat, C., 1967, Modified Gardner delay time and constant distance correlation interpretation, *in* Seismic refraction prospecting: A. W. Musgrave, Ed., SEG, Tulsa, p. 171–193.

Levin, F. K., and Ingram, J. D., 1960, Head waves from a bed of finite thickness: Geophysics, v. 27, p. 753–765.

McPhail, M. R., 1967, The midpoint method of interpreting a refraction survey: Seismic refraction prospecting, A. W. Musgrave, Ed., SEG, Tulsa, p. 260–266.

Maillet, R., and Bazerque, J., 1931, La prospection seismique du sous-sol: Annales des Mines, v. 20, p. 314.

Meidev, T., 1960, Nomograms to speed up seismic refraction computations: Geophysics, v. 25, p. 1035–1053.

Merrick, N. P., Odins, J. A., and Greenhalgh, S. A., 1978, A blind zone solution to the problem of hidden layers within sequence of horizontal or dipping refractors: Geophys. Prosp., v. 26, p. 703–721.

Mota, L., 1954, Determination of dip and depths of geological layers by the seismic refraction method: Geophysics, v. 19, p. 242–254.

Musgrave, A. W., Ed., 1967, Seismic refraction prospecting: SEG, Tulsa.

Musgrave, A. W., and Bratton, R. H., 1967, Practical application of Blondeau weathering solution: Seismic refraction prospecting, A. W. Musgrave, Ed., SEG, Tulsa, p. 231–246.

Northwood, E. J., 1967, Nomograms for curved-ray problems in overburden: in Seismic refraction prospecting, A. W. Musgrave, Ed., SEG, Tulsa, p. 296–303.

Olhovich, V. A., 1959, Curso de Sismologia Applicada: Mexico, D. F., Editorial Reverte, S.A.

Pakiser, L. C., and Black, R. A., 1957, Exploration for ancient channels with the exploration seismograph: Geophysics, v. 22, p. 32–47.

Palmer, D., 1974, An application of the time section in shallow seismic refraction studies: M.Sc. thesis, Univ. of Sydney, 157 p., unpublished.

———1979, What is the future for seismic refraction methods?: Bull. Austral. SEG, v. 10, p. 215–217.

Paterson, N. R., 1956, Seismic wave propagation in porous granular media: Geophysics, v. 21, p. 691–714.

Press, F., Oliver, J., and Ewing., M., 1954, Seismic model study of refractions from a layer of finite thickness: Geophysics, v. 19, p. 338–401.

Peraldi, R., and Clement, A., 1972, Digital processing of refraction data—Study of first arrivals: Geophys. Prosp., v. 20, p. 529–548.

Puzyrez, N. N., 1972, On conditions for omission of layers during registration of first arrivals: Inst. Geol. and Geophys., Siberian Branch, Acad. Sci., Ukranian SSR, Geophys. Trans., v. 48, p. 17–30.

Rockwell, D. W., 1967, A general wavefront method: Seismic refraction prospecting, in A. W. Musgrave, Ed., SEG, Tulsa, p. 363–415.

Scott, J. H., 1973, Seismic refraction modeling by computer: Geophysics, v. 38, p. 271–284.

Sendlein, L. V. A., 1968, Discussion on "Refraction determination of water table depth and alluvium thickness" (Duguid, Geophysics, v. 33, p. 481–488): Geophysics, v. 33, p. 1019–1021.

Sheriff, R. E., 1973, Encyclopedic dictionary of exploration geophysics: SEG, Tulsa.

Shima, E., 1957, Note on the depth calculation by the seismic refraction method: J. SEG Japan, v. 10, p. 16–18.

Singh, S., 1978, An iterative method for detailed depth determination from refraction for an uneven interface: Geophys. Prosp., v. 26, p. 303–311.

Slichter, L. B., 1932, The theory of the interpretation of seismic traveltime curves in horizontal structures: Physics, v. 3, p. 273–295.

Slotnick, M. M., 1950, A graphical method for the interpretation of refraction profile data: Geophysics, v. 15, p. 163–180.

Soske, J. L., 1959, The blind zone problem in engineering geophysics: Geophysics, v. 24, p. 359–365.

Stulken, E. J., 1967, Constructions, graphs and nomograms for refraction computations: in Seismic refraction prospecting, A. W. Musgrave, Ed., SEG, Tulsa, p. 304–329.

Tarrant, L. H., 1956, A rapid method of determining the form of a seismic refractor from line profile results: Geophys. Prosp., v. 4, p. 131–139.

Telford, W. M., Geldard, L. P., Sheriff, R. E., and Keys, D. A., 1976, Applied geophysics, New York, Cambridge University Press.

Thornburgh, H. R., 1930, Wavefront diagrams in seismic interpretation: AAPG Bull., v. 14, p. 185–200.

Woolley, W. C., Musgrave, A. W., and Gray, H., 1967, A method of in-line refraction profiling: Seismic refraction prospecting, A. W. Musgrave, Ed., SEG, Tulsa, p. 267–289.

Wyrobek, S. M., 1956, Application of delay and intercept times in the interpretation of multilayer refraction time distance curves: Geophys. Prosp., v. 4, p. 112–130.

Appendix A
Synthetic Models

Distance (m)	Traveltime (msec)							
	Flat topography irregular refractor		Irregular topography flat refractor		Irregular topography irregular refractor		Flat topography irregular refractor linear velocity in overburden	
0		99.0	30.9	110.1		99.0		100.4
5		97.0	32.2	108.9		97.0		99.2
10		96.0	33.4	107.6		96.0		98.0
15	23.0	95.0	34.7	106.4		94.8	24.8	96.8
20	24.3	93.8	35.9	105.1	23.8	94.2	26.1	95.6
25	25.4	92.5	37.2	103.9	25.1	92.6	27.3	94.5
30	26.5	91.5	38.4	102.6	26.4	92.0	28.6	93.4
35	28.0	90.5	39.7	101.4	27.9	90.8	29.8	92.4
40	29.0	89.5	40.9	101.1	29.1	90.2	31.1	91.6
45	30.4	89.0	42.2	98.9	30.6	89.0	32.3	91.3
50	32.0	90.0	43.4	97.6	31.9	90.0	33.6	92.0
55	33.0	91.0	45.3	97.0	34.0	92.0	34.8	90.8
60	34.8	90.0	47.1	96.3	36.0	92.0	36.4	89.5
65	37.0	88.3	49.0	95.7	39.1	91.0	38.7	88.3
70	39.8	87.0	50.9	95.1	42.1	90.0	41.3	87.0
75	42.5	85.7	52.7	94.4	44.8	89.0	44.1	85.8
80	45.4	84.3	53.4	92.6	47.3	87.0	45.7	84.5
85	47.0	82.6	54.0	90.7	49.0	84.0	46.9	82.9
90	48.7	80.3	54.6	88.8	49.7	82.0	48.2	81.1
95	49.2	78.2	55.3	87.0	50.4	79.0	49.4	79.4
100	50.1	76.2	55.9	85.1	51.0	76.0	50.7	77.6
105	50.5	74.0	57.2	83.9	51.5	74.0	51.7	75.8
110	50.8	72.0	58.4	82.6	51.5	72.0	52.2	74.0
115	51.1	70.0	59.7	81.4	51.5	70.0	52.7	72.3
120	51.5	68.7	60.9	80.1	52.0	69.0	53.2	70.7
125	51.8	68.5	59.1	75.8	49.2	65.0	53.6	70.0
130	52.5	68.2	60.3	74.5	49.6	65.0	54.2	69.6
135	54.0	68.0	61.6	73.3	51.3	65.2	55.8	69.1
140	56.0	67.5	62.8	72.0	53.3	64.5	57.6	68.6
145	58.0	67.2	64.1	70.8	55.2	64.5	59.4	68.1
150	60.0	67.1	65.3	69.5	57.9	64.0	61.2	67.1
155	62.0	66.0	67.2	68.9	60.4	63.6	63.0	65.9
160	64.0	64.5	69.1	68.3	63.0	63.1	64.8	64.7
165	66.0	63.5	70.9	67.6	65.5	62.7	66.5	63.7
170	68.0	62.5	72.8	67.2	68.0	62.1	68.2	62.7
175	69.2	61.4	74.7	66.8	69.0	61.8	69.4	61.7
180	70.4	60.4	76.5	66.4	71.0	61.2	70.7	60.8
185	71.6	59.2	78.4	66.1	73.0	61.0	71.8	59.8
190	72.6	58.5	80.0	65.7	74.5	61.0	72.8	58.9
195	74.0	58.5	81.6	65.3	76.0	61.5	73.8	58.6
200	74.6	58.5	83.3	65.0	78.2	62.0	74.8	58.2
205	75.9	58.5	83.6	63.3	78.8	61.0	75.8	57.8
210	77.1	58.5	84.0	61.7	79.2	61.0	77.1	57.4
215	79.0	58.5	84.4	60.1	81.0	60.0	78.7	56.2

Appendix A
Synthetic Models (continued)

Distance (m)	Traveltime (msec)							
	Flat topography irregular refractor		Irregular topography flat refractor		Irregular topography irregular refractor		Flat topography irregular refractor linear velocity in overburden	
220	81.0	56.3	84.7	58.4	82.0	57.0	80.2	54.8
225	83.0	54.4	85.1	56.8	83.0	54.0	81.6	53.3
230	84.8	52.5	86.1	55.8	85.0	52.1	83.1	51.7
235	87.0	50.3	87.1	54.8	87.0	50.1	84.6	50.2
240	87.0	48.7	88.1	53.8	87.0	49.0	85.8	48.7
245	87.0	47.3	89.1	52.8	87.0	47.8	86.2	47.3
250	87.0	46.2	90.1	51.8	87.0	46.1	86.5	46.3
255	87.0	45.0	94.3	54.0	90.0	48.0	86.9	45.3
260	87.2	43.5	95.3	53.0	90.7	47.0	87.3	44.3
265	88.0	42.4	96.3	52.0	91.2	45.0	88.1	44.1
270	89.0	40.5	97.3	51.0	92.2	43.0	89.1	41.5
275	90.0	38.5	98.3	50.0	93.0	41.8	90.1	39.9
280	90.2	36.5	98.6	48.3	93.0	39.2	91.0	38.2
285	90.2	35.0	99.0	46.7	92.8	37.0	91.4	36.5
290	90.2	33.0	99.4	45.1	92.2	34.0	91.6	34.8
295	90.2	31.0	99.7	43.4	91.3	32.0	91.9	33.1
300	90.2	30.0	100.1	41.8	91.0	30.0	92.0	31.6
305	90.2	28.5	101.1	40.8	91.0	28.7	92.2	30.6
310	91.0	27.5	102.1	39.8	91.0	27.0	92.7	29.6
315	92.0	26.5	103.1	38.8	92.0	26.0	93.7	28.6
320	92.5	25.3	104.1	37.8	92.8	25.0	94.7	27.6
325	93.5	24.5	105.1	36.8	93.8	24.0	95.6	26.6
330	94.3	23.3	106.1	35.8	94.8	23.0	96.6	25.6
335	95.2	22.5	107.1	34.8	95.8		97.6	24.6
340	96.3		108.1	33.8	96.8		98.6	
345	97.0		109.1	32.8	97.8		99.6	
350	99.0		110.1	31.8	98.8		100.5	

Appendix B
Traveltimes and locations from Welcome Reef dam site investigation. Station spacing (location)=10 m. Time in msec.

Line no. 3

SHOT NUMBER 61 LOCATION 96

LOCATION	TIME	LOCATION	TIME	LOCATION	TIME	LOCATION	TIME
96		96.5	18	97	23	97.5	32
98	36	98.5	43	99	50	99.5	56
100	64	100.5	69	101	72	101.5	75
102	76	102.5	80	103	81	103.5	82
104	84	104.5	86	105	87	105.5	89
106	92	106.5	94	107	95	107.5	97
108	100						

SHOT NUMBER 62 LOCATION 102

LOCATION	TIME	LOCATION	TIME	LOCATION	TIME	LOCATION	TIME
96	77	96.5	73	97	71	97.5	65
98	59	98.5	52	99	45	99.5	40
100	33	100.5	29	101	23	101.5	16
102		102.5	16	103	24	103.5	31
104	38	104.5	44	105	51	105.5	54
106	57	106.5	60	107	63	107.5	65

SHOT NUMBER 63 LOCATION 108

LOCATION	TIME	LOCATION	TIME	LOCATION	TIME	LOCATION	TIME
96	100	96.5	97	97	96	97.5	94
98	91	98.5	88	99	85	99.5	82
100	79	100.5	77	101	74	101.5	71
102	68	102.5	68	103	63	104	57
104.5	54	105	50	105.5	42	106	36
106.5	28	107	20	107.5	13	108	

SHOT NUMBER 64 LOCATION 120

LOCATION	TIME	LOCATION	TIME	LOCATION	TIME	LOCATION	TIME
96	138	96.5	136	97	136	97.5	134
98	133	98.5	131	99	128	99.5	127
100	125	100.5	123	101	121	101.5	118
102	116	102.5	116	103	113	103.5	112
104	109	104.5	107	105	104	105.5	102
106	101	106.5	99	107	98	107.5	95

SHOT NUMBER 65 LOCATION 120

LOCATION	TIME	LOCATION	TIME	LOCATION	TIME	LOCATION	TIME
108	93	108.5	92	109	90	109.5	87
110	86	110.5	83	111	80	111.5	80
112	76	112.5	75	113	73	113.5	71
114	68	114.5	65	115	64	115.5	60
116	58	116.5	55	117	49	117.5	41
118	34	118.5	29	119	24	119.5	19
120							

SHOT NUMBER 66 LOCATION 114

LOCATION	TIME	LOCATION	TIME	LOCATION	TIME	LOCATION	TIME
108	66	108.5	65	109	62	109.5	59
110	58	110.5	55	111	51	111.5	45
112	35	112.5	30	113	23	113.5	18
114		114.5	16	115	23	115.5	28
116	35	116.5	42	117	49	117.5	56
118	57	118.5	60	119	65	119.5	68
120	68						

SHOT NUMBER 67 LOCATION 108

LOCATION	TIME	LOCATION	TIME	LOCATION	TIME	LOCATION	TIME
108		108.5	15	109	20	109.5	25
110	31	110.5	39	111	47	111.5	54
112	55	112.5	57	113	59	113.5	61
114	65	114.5	66	115	70	115.5	72
116	74	116.5	76	117	79	117.5	79
118	81	118.5	85	119	89	119.5	91
120	91						

SHOT NUMBER 68 LOCATION 96

LOCATION	TIME	LOCATION	TIME	LOCATION	TIME	LOCATION	TIME
108.5	102	109	104	109.5	105	110	107
110.5	108	111	109	111.5	111	112	110
112.5	111	113	114	113.5	116	114	118
114.5	119	115	123	115.5	124	116	126
116.5	127	117	129	117.5	130	118	131
118.5	133	119	137	119.5	139	120	138

SHOT NUMBER 69 LOCATION 84

LOCATION	TIME	LOCATION	TIME	LOCATION	TIME	LOCATION	TIME
97	112	98	119	99	120	100	122
101	123	102	125	103	127	104	129
105	130	106	133	107	134	108	138
109	138	110	141	111	142	112	143
113	147	114	152	115	156	116	158
117	160	118	161	119	167	120	168

SHOT NUMBER 70 LOCATION 72

LOCATION	TIME	LOCATION	TIME	LOCATION	TIME	LOCATION	TIME
97	147	98	150	99	150	100	153
101	154	102	155	103	156	104	158
105	159	106	160	107	163	108	167
109	166	110	168	111	170	112	172
113	175	114	179	115	184	116	186
117	187	118	189	119	195	120	196

SHOT NUMBER 71 LOCATION 132

LOCATION	TIME	LOCATION	TIME	LOCATION	TIME	LOCATION	TIME
96	205	97	202	98	199	99	196
100	193	101	188	102	184	103	182
104	181	105	176	106	174	107	171
108	169	109	165	110	163	111	159
112	155	113	153	114	150	115	146
116	141	117	137	118	133	119	133

SHOT NUMBER 72 LOCATION 144

LOCATION	TIME	LOCATION	TIME	LOCATION	TIME	LOCATION	TIME
96	238	97	234	98	233	99	288
100	225	101	221	102	218	103	216
104	213	105	211	106	208	107	206
108	204	109	200	110	199	111	196
112	194	113	193	114	192	115	189
116	185	117	184	118	183	119	185

SHOT NUMBER 73 LOCATION 96

LOCATION	TIME	LOCATION	TIME	LOCATION	TIME	LOCATION	TIME
121	141	122	146	123	152	124	160
125	165	126	170	127	176	128	185
129	192	130	195	131	200	132	204
133	209	134	210	135	213	136	216
137	219	138	221	139	226	140	227
141	229	142	235	143	235	144	239

SHOT NUMBER 74 LOCATION 108

LOCATION	TIME	LOCATION	TIME	LOCATION	TIME	LOCATION	TIME
121	96	122	103	123	110	124	118
125	124	126	129	127	136	128	146
129	152	130	157	131	164	132	168
133	174	134	175	135	179	136	181
137	185	138	187	139	191	140	192
141	195	142	201	143	201	144	204

SHOT NUMBER 75 LOCATION 156

LOCATION	TIME	LOCATION	TIME	LOCATION	TIME	LOCATION	TIME
120	225	121	227	122	230	123	231
124	230	125	229	126	227	127	224
128	223	129	219	130	215	131	214
132	212	133	210	134	205	135	204
136	198	137	191	138	184	139	177
140	173	141	169	142	162	143	155

SHOT NUMBER 76 LOCATION 168

LOCATION	TIME	LOCATION	TIME	LOCATION	TIME	LOCATION	TIME
120	258	121	262	122	262	123	265
124	265	125	263	126	263	127	260
128	258	129	257	130	252	131	251
132	250	133	248	134	243	135	243
136	242	137	237	138	234	139	231
140	227	141	227	142	225	143	221

SHOT NUMBER 77 LOCATION 120

LOCATION	TIME	LOCATION	TIME	LOCATION	TIME	LOCATION	TIME
120		120.5	16	121	22	121.5	27
122	33	122.5	40	123	45	123.5	50
124	57	124.5	63	125	69	125.5	74
126	82	126.5	86	127	91	127.5	98
128	101	128.5	106	129	112	129.5	113
130	117	130.5	120	131	127	131.5	126
132	131						

SHOT NUMBER 78 LOCATION 126

LOCATION	TIME	LOCATION	TIME	LOCATION	TIME	LOCATION	TIME
120	84	120	78	121	72	121.5	66
122	61	122.5	55	123	43	123.5	35
124	28	124.5	22	125	17	125.5	13
126		126.5	13	127	18	127.5	24
128	31	128.5	37	129	43	129.5	50
130	54	130.5	61	131	69	131.5	72
132	81						

SHOT NUMBER 121 LOCATION 132

LOCATION	TIME	LOCATION	TIME	LOCATION	TIME	LOCATION	TIME
120	131	120.5	130	121	128	121.5	128
122	128	122.5	122	123	120	123.5	119
124	112	124.5	103	125	98	125.5	92
126	82	126.5	77	127	70	127.5	66
128	59	128.5	53	129	48	129.5	41
130	33	130.5	29	131	21	131.5	12
132							

SHOT NUMBER 80 LOCATION 144

LOCATION	TIME	LOCATION	TIME	LOCATION	TIME	LOCATION	TIME
120	188	120.5	187	121	188	121.5	189
122	192	122.5	194	123	191	123.5	189
124	187	124.5	186	125	183	125.5	183
126	180	126.5	177	127	173	127.5	174
128	167	128.5	163	129	164	129.5	160
130	156	130.5	156	131	155	131.5	148

SHOT NUMBER 81 LOCATION 120

LOCATION	TIME	LOCATION	TIME	LOCATION	TIME	LOCATION	TIME
132.5	133	133	137	133.5	140	134	143
134.5	147	135	150	135.5	152	136	155
136.5	155	137	161	137.5	162	138	167
138.5	170	139	172	139.5	173	140	174
140.5	177	141	179	141.5	182	142	183
142.5	182	143	184	143.5	185	144	187

SHOT NUMBER 82 LOCATION 132

LOCATION	TIME	LOCATION	TIME	LOCATION	TIME	LOCATION	TIME
132		132.5	21	133	26	133.5	32
134	38	134.5	46	135	53	135.5	60
136	69	136.5	76	137	86	137.5	91
138	99	138.5	106	139	112	139.5	119
140	125	140.5	129	141	132	141.5	135
142	139	142.5	141	143	142	143.5	147
144	151						

SHOT NUMBER 83 LOCATION 138

LOCATION	TIME	LOCATION	TIME	LOCATION	TIME	LOCATION	TIME
132	105	132.5	93	133	85	133.5	77
134	69	134.5	61	135	52	135.5	46
136	37	136.5	29	137	21	137.5	16
138		138.5	15	139	21	139.5	29
140	37	140.5	47	141	57	141.5	64
142	68	142.5	76	143	86	143.5	94
144	100						

SHOT NUMBER 84 LOCATION 144

LOCATION	TIME	LOCATION	TIME	LOCATION	TIME	LOCATION	TIME
132	154	132.5	149	133	145	133.5	142
134	140	134.5	139	135	134	135.5	131
136	127	136.5	120	137	113	137.5	110
138	103	138.5	93	139	84	139.5	75
140	67	140.5	60	141	52	141.5	46
142	36	142.5	28	143	22	143.5	17
144							

SHOT NUMBER 97 LOCATION 156

LOCATION	TIME	LOCATION	TIME	LOCATION	TIME	LOCATION	TIME
169	161	170	164	171	169	172	171
173	173	174	179	175	186	176	192
177	196	178	200	179	203	180	206
181	208	182	212	183	217	184	220
185	222	186	223	187	230	188	231
189	236	190	241	191	247	192	249

SHOT NUMBER 98 LOCATION 144

LOCATION	TIME	LOCATION	TIME	LOCATION	TIME	LOCATION	TIME
169	217	170	218	171	220	172	222
173	223	174	226	175	232	176	234
177	239	178	241	179	245	180	246
181	248	182	250	183	255	184	257
185	258	186	259	187	265	188	268
189	271	190	275	191	279	192	282

SHOT NUMBER 99 LOCATION 204

LOCATION	TIME	LOCATION	TIME	LOCATION	TIME	LOCATION	TIME
168	251	169	246	170	242	171	238
172	232	173	228	174	225	175	223
176	218	177	216	178	214	179	209
180	205	181	203	182	200	183	196
184	189	185	186	186	182	187	180
188	174	189	171	190	168	191	161

SHOT NUMBER 100 LOCATION 216

LOCATION	TIME	LOCATION	TIME	LOCATION	TIME	LOCATION	TIME
168	273	169	269	170	264	171	263
172	258	173	254	174	253	175	248
176	245	177	245	178	241	179	237
180	234	181	232	182	229	183	227
184	221	185	217	186	215	187	213
188	207	189	205	190	204	191	200

SHOT NUMBER 101 LOCATION 168

LOCATION	TIME	LOCATION	TIME	LOCATION	TIME	LOCATION	TIME
168		168.5	14	169	19	169.5	24
170	28	170.5	33	171	40	171.5	45
172	51	172.5	57	173	64	173.5	71
174	76	174.5	84	175	90	175.5	95
176	102	176.5	106	177	115	177.5	121
178	130	178.5	141	179	150	179.5	151
180	154						

SHOT NUMBER 102 LOCATION 174

LOCATION	TIME	LOCATION	TIME	LOCATION	TIME	LOCATION	TIME
168	78	169	65	169.5	60	170	53
170.5	48	171	41	171.5	35	172	29
172.5	22	173	14	173.5	9	174	
174.5	10	175	17	175.5	22	176	29
176.5	36	177	43	177.5	50	178	56
178.5	62	179	68	179.5	73	180	83

SHOT NUMBER 103 LOCATION 180

LOCATION	TIME	LOCATION	TIME	LOCATION	TIME	LOCATION	TIME
168	154	168.5	152	169	148	169.5	144
170	137	170.5	128	171	122	171.5	114
172	105	172.5	102	173	94	173.5	90
174	85	174.5	77	175	70	175.5	63
176	54	176.5	46	177	39	177.5	31
178	24	178.5	20	179	16	179.5	13
180							

SHOT NUMBER 104 LOCATION 192

LOCATION	TIME	LOCATION	TIME	LOCATION	TIME	LOCATION	TIME
168	208	168.5	206	169	205	169.5	202
170	199	170.5	197	171	192	171.5	189
172	188	172.5	186	173	182	173.5	181
174	180	174.5	178	175	176	175.5	172
176	170	176.5	168	177	168	177.5	166
178	166	178.5	162	179	159	179.5	157

SHOT NUMBER 105 LOCATION 192

LOCATION	TIME	LOCATION	TIME	LOCATION	TIME	LOCATION	TIME
180	156	180.5	153	181	150	181.5	147
182	139	182.5	130	183	121	183.5	112
184	102	184.5	98	185	95	185.5	87
186	78	186.5	71	187	65	187.5	56
188	47	188.5	40	189	36	189.5	28
190	21	190.5	17	191	11	191.5	7
192							

SHOT NUMBER 106 LOCATION 186

LOCATION	TIME	LOCATION	TIME	LOCATION	TIME	LOCATION	TIME
180	96	180.5	85	181	83	181.5	76
182	56	182.5	49	183	44	183.5	37
184	30	184.5	27	185	22	185.5	12
186		186.5	11	187	19	187.5	27
188	31	188.5	37	189	44	189.5	49
190	56	190.5	63	191	67	191.5	76
192	81						

SHOT NUMBER 107 LOCATION 180

LOCATION	TIME	LOCATION	TIME	LOCATION	TIME	LOCATION	TIME
180		180.5	8	181	15	181.5	19
182	24	182.5	33	183	37	183.5	43
184	53	184.5	56	185	64	185.5	77
186	88	186.5	90	187	98	187.5	111
188	144	188.5	120	189	128	189.5	135
190	138	190.5	145	191	149	191.5	154
192	147						

SHOT NUMBER 108 LOCATION 168

LOCATION	TIME	LOCATION	TIME	LOCATION	TIME	LOCATION	TIME
180.5	156	181	159	181.5	163	182	164
182.5	168	183	170	183.5	174	184	175
184.5	175	185	177	185.5	178	186	179
186.5	180	187	184	187.5	187	188	187
188.5	189	189	193	189.5	195	190	199
190.5	202	191	205	191.5	207	192	208

SHOT NUMBER 109 LOCATION 180

LOCATION	TIME	LOCATION	TIME	LOCATION	TIME	LOCATION	TIME
194	164	195	166	196	170	197	174
198	180	199	185	200	188	201	192
202	197	203	201	204	205	205	207
206	209	207	212	208	216	209	219
210	221	211	222	212	225	213	229
214	230	215	233	216	235		

SHOT NUMBER 110 LOCATION 168

LOCATION	TIME	LOCATION	TIME	LOCATION	TIME	LOCATION	TIME
193	212	194	215	195	217	196	221
197	225	198	231	199	233	200	235
201	240	202	243	203	247	204	251
205	253	206	253	207	256	208	257
209	261	210	263	211	264	212	266
213	268	214	272	215	272	216	274

SHOT NUMBER 111 LOCATION 240

LOCATION	TIME	LOCATION	TIME	LOCATION	TIME	LOCATION	TIME
192	273	193	272	194	268	195	266
196	262	197	260	198	259	199	254
200	252	201	250	202	247	203	243
204	238	205	234	206	231	207	228
208	224	209	223	210	218	211	213
212	210	213	208	214	204	215	200

SHOT NUMBER 112 LOCATION 288

LOCATION	TIME	LOCATION	TIME	LOCATION	TIME	LOCATION	TIME
193	236	194	235	195	231	196	228
197	227	198	224	199	219	200	217
201	214	202	210	203	205	204	200
205	195	206	190	207	187	208	183
209	181	210	175	211	169	212	165
213	159	214	155	215	149	216	145

SHOT NUMBER 113 LOCATION 216

LOCATION	TIME	LOCATION	TIME	LOCATION	TIME	LOCATION	TIME
204	144	204.5	143	205	140	205.5	137
206	135	206.5	133	207	130	207.5	125
208	118	208.5	113	209	108	209.5	101
210	71	210.5	65	211	58	211.5	55
212	48	212.5	43	213	36	213.5	31
214	24	214.5	19	215	14	215.5	11
216							

SHOT NUMBER 114 LOCATION 210

LOCATION	TIME	LOCATION	TIME	LOCATION	TIME	LOCATION	TIME
204	76	204.5	67	205	61	205.5	54
206	47	206.5	42	207	33	207.5	29
208	24	208.5	19	209	13	209.5	8
210		210.5	7	211	13	211.5	18
212	23	212.5	28	213	34	213.5	39
214	46	214.5	53	215	56	215.5	64
216	71						

SHOT NUMBER 115 LOCATION 204

LOCATION	TIME	LOCATION	TIME	LOCATION	TIME	LOCATION	TIME
204		204.5	10	205	15	205.5	19
206	25	206.5	30	207	36	207.5	42
208	49	208.5	57	209	62	209.5	69
210	78	210.5	84	211	91	211.5	115
212	120	212.5	127	213	134	213.5	137
214	140	214.5	141	215	142	215.5	145
216	146						

SHOT NUMBER 116 LOCATION 192

LOCATION	TIME	LOCATION	TIME	LOCATION	TIME	LOCATION	TIME
204.5	158	205	161	205.5	162	206	162
206.5	162	207	166	207.5	167	208	171
208.5	173	209	174	209.5	176	210	178
210.5	180	211	180	211.5	183	212	184
212.5	186	213	188	213.5	191	214	192
214.5	192	215	193	215.5	195	216	196

SHOT NUMBER 117 LOCATION 192

LOCATION	TIME	LOCATION	TIME	LOCATION	TIME	LOCATION	TIME
192		192.5	9	193	14	193.5	20
194	25	194.5	33	195	39	195.5	45
196	51	196.5	57	197	65	197.5	70
198	78	198.5	86	199	93	199.5	100
200	107	200.5	113	201	119	201.5	129
202	135	202.5	145	203	153	203.5	156
204	158						

SHOT NUMBER 118 LOCATION 198

LOCATION	TIME	LOCATION	TIME	LOCATION	TIME	LOCATION	TIME
192	78	192.5	71	193	66	193.5	59
194	51	194.5	45	195	38	195.5	33
196	26	196.5	22	197	14	197.5	10
198		198.5	11	199	18	199.5	23
200	28	200.5	33	201	39	201.5	45
202	51	202.5	57	203	64	203.5	73
204	77						

SHOT NUMBER 119 LOCATION 204

LOCATION	TIME	LOCATION	TIME	LOCATION	TIME	LOCATION	TIME
192	157	192.5	154	193	152	193.5	144
194	136	194.5	126	195	116	195.5	112
196	103	196.5	98	197	89	197.5	83
198	76	198.5	68	199	62	199.5	55
200	48	200.5	42	201	37	201.5	31
202	24	202.5	18	203	13	203.5	9
204							

SHOT NUMBER 120 LOCATION 216

LOCATION	TIME	LOCATION	TIME	LOCATION	TIME	LOCATION	TIME
192	197	192.5	194	193	192	193.5	191
194	188	194.5	186	195	183	195.5	182
196	182	196.5	180	197	178	197.5	177
198	174	198.5	172	199	170	199.5	167
200	165	200.5	162	201	161	201.5	157
202	154	202.5	152	203	150	203.5	148

Line no. 4

SHOT NUMBER 189 LOCATION 96

LOCATION	TIME	LOCATION	TIME	LOCATION	TIME	LOCATION	TIME
49	271	50	272	51	271	52	268
53	265	54	261	55	259	56	258
57	255	58	251	59	249	60	249
61	248	62	245	63	244	64	241
65	237	66	234	67	230	68	227
69	227	70	220	71	215	72	210

SHOT NUMBER 190 LOCATION 84

LOCATION	TIME	LOCATION	TIME	LOCATION	TIME	LOCATION	TIME
49	231	50	232	51	229	52	227
53	223	54	220	55	218	56	214
57	212	58	208	59	205	60	203
61	198	62	197	63	189	64	181
65	175	66	172	67	166	68	158
69	153	70	148	71	140		

SHOT NUMBER 191 LOCATION 72

LOCATION	TIME	LOCATION	TIME	LOCATION	TIME	LOCATION	TIME
48	193	49	196	50	194	51	189
52	181	53	176	54	171	55	165
56	160	57	156	58	150	59	142
60	139	61	133	62	112	63	100
64	84	65	73	66	64	67	52
68	43	69	33	70	23	71	13
72							

SHOT NUMBER 192 LOCATION 66

LOCATION	TIME	LOCATION	TIME	LOCATION	TIME	LOCATION	TIME
49	164	50	159	51	152	52	145
53	138	54	134	55	118	56	104
57	95	58	85	59	74	60	63
61	52	62	42	63	31	64	22
65	13	66		67	14	68	25
69	35	70	43	71	54	72	65

SHOT NUMBER 193 LOCATION 60

LOCATION	TIME	LOCATION	TIME	LOCATION	TIME	LOCATION	TIME
48	133	49	122	50	109	51	100
52	88	53	76	54	66	55	55
56	44	57	35	58	27	59	18
60	46	61	18	62	28	63	36
64	46	65	56	67	78	68	88
69	100	70	109	71	132	72	140

SHOT NUMBER 194 LOCATION 54

LOCATION	TIME	LOCATION	TIME	LOCATION	TIME	LOCATION	TIME
48	67	49	58	50	48	51	38
52	28	53	16	54		55	17
56	26	57	37	58	46	59	57
6	68	61	77	62	89	63	100
64	110	65	122	66	133	67	145
68	149	69	154	70	160	71	166

SHOT NUMBER 195 LOCATION 48

LOCATION	TIME	LOCATION	TIME	LOCATION	TIME	LOCATION	TIME
48		49	16	50	25	51	36
52	47	53	56	54	67	55	76
56	86	57	98	58	109	59	122
60	131	61	139	62	146	63	152
64	161	65	165	66	172	67	176
68	182	69	186	70	189	71	193
72	195						

SHOT NUMBER 196 LOCATION 36

LOCATION	TIME	LOCATION	TIME	LOCATION	TIME	LOCATION	TIME
49	96	50	104	51	110	52	119
53	128	54	139	55	144	56	151
57	152	58	161	59	166	60	170
61	173	62	175	63	180	64	182
65	186	66	191	67	196	68	196
69	196	70	199	71	202	72	203

SHOT NUMBER 197 LOCATION 24

LOCATION	TIME	LOCATION	TIME	LOCATION	TIME	LOCATION	TIME
48	121	49	125	50	131	51	140
52	147	53	153	54	163	55	169
56	172	57	179	58	184	59	188
60	190	61	195	62	196	63	199
64	202	65	206	66	209	67	211
68	212	69	218	70	218	71	223

SHOT NUMBER 198 LOCATION 72

LOCATION	TIME	LOCATION	TIME	LOCATION	TIME	LOCATION	TIME
25	224	26	222	27	218	28	216
29	214	30	211	31	210	32	210
33	208	34	206	35	205	36	204
37	202	38	200	39	196	40	195
41	194	42	191	43	190	44	190
45	187	46	190	47	194	48	191

SHOT NUMBER 199 LOCATION 60

LOCATION	TIME	LOCATION	TIME	LOCATION	TIME	LOCATION	TIME
24	194	25	189	26	189	27	186
28	184	29	183	30	177	31	175
32	174	33	171	34	168	35	165
36	166	37	163	38	160	39	155
40	150	41	147	42	145	43	141
44	140	45	135	46	134	47	133

SHOT NUMBER 200 LOCATION 48

LOCATION	TIME	LOCATION	TIME	LOCATION	TIME	LOCATION	TIME
24	121	25	119	26	115	27	112
28	110	29	108	30	102	31	101
32	99	33	98	34	92	35	90
36	89	37	86	38	78	39	77
40	74	41	69	42	64	43	57
44	48	45	37	46	27	47	18
48							

SHOT NUMBER 201 LOCATION 42

LOCATION	TIME	LOCATION	TIME	LOCATION	TIME	LOCATION	TIME
25	106	26	102	27	100	28	98
29	94	30	89	31	86	32	86
33	80	34	77	35	72	36	69
37	63	38	58	39	42	40	31
41	18	42		43	20	44	30
45	39	46	51	47	59	48	67

SHOT NUMBER 202 LOCATION 36

LOCATION	TIME	LOCATION	TIME	LOCATION	TIME	LOCATION	TIME
24	94	25	88	26	81	27	79
28	76	29	71	30	66	31	61
32	56	33	45	34	30	35	20
36		37	21	38	39	39	49
40	61	41	66	42	69	43	71
44	75	45	76	46	80	47	87
48	88						

SHOT NUMBER 203 LOCATION 30

LOCATION	TIME	LOCATION	TIME	LOCATION	TIME	LOCATION	TIME
24	62	25	54	26	50	27	41
28	31	29	21	30		31	22
32	41	33	53	34	59	35	66
36	69	37	75	38	81	39	83
40	87	41	89	42	91	43	91
44	94	45	92	46	99	47	103

SHOT NUMBER 204 LOCATION 24

LOCATION	TIME	LOCATION	TIME	LOCATION	TIME	LOCATION	TIME
24		25	19	26	31	27	43
28	48	29	56	30	63	31	67
32	75	33	80	34	83	35	89
36	92	37	97	38	102	39	102
40	108	41	107	42	110	43	109
44	112	45	112	46	115	47	119

SHOT NUMBER 205 LOCATION 12

LOCATION	TIME	LOCATION	TIME	LOCATION	TIME	LOCATION	TIME
25	91	26	94	27	100	28	102
29	106	30	108	31	109	32	114
33	119	34	123	35	127	36	131
37	132	38	134	39	133	40	138
41	138	42	140	43	140	44	141
45	141	46	144	47	150	48	155

SHOT NUMBER 206 LOCATION 0

LOCATION	TIME	LOCATION	TIME	LOCATION	TIME	LOCATION	TIME
24	133	25	136	26	136	27	141
28	144	29	145	30	147	31	148
32	155	33	159	34	160	35	164
36	167	37	167	38	168	39	171
40	173	41	175	42	175	43	174
44	175	45	177	46	181	47	185

SHOT NUMBER 7 LOCATION 54

LOCATION	TIME	LOCATION	TIME	LOCATION	TIME	LOCATION	TIME
62	109	63	121	64	128	65	135
66	140	67	145				

SHOT NUMBER 8 LOCATION 60

LOCATION	TIME	LOCATION	TIME	LOCATION	TIME	LOCATION	TIME
68	113	69	123	70	128	71	132

SHOT NUMBER 9 LOCATION 66

LOCATION	TIME	LOCATION	TIME	LOCATION	TIME	LOCATION	TIME
54	134	55	130	56	122		

SHOT NUMBER 10 LOCATION 72

LOCATION	TIME	LOCATION	TIME	LOCATION	TIME	LOCATION	TIME
61	133	62	121	63	113		

Index

Accuracy of ray tracing 54
Accurate measurement of arrival times 59
Ambiguity 56, 59
  field procedures for 37
  traveltime curves 37
Apparent refractor velocity 9, 13
Average velocity 41–47, 74, 83
  comparison with hidden layer errors 44–45
  three-layer case 43–44
Average velocity method 46
  test of 46

Blind zone 37–38

Calculation of an expected *XY* value 33
Complex velocity-depth profile 38
Consistency between data and interpretation 54
Continuous change of velocity with depth 15–16, 29
Conventional intercept time 1
Conventional intercept time method 1
Conventional reciprocal method 1
Conventional time-depth 13
Corrections for surface layers 45–46
Critical distance method 1

Data processing 55–56
Delay time, the 13
Delay time method 1
Depth conversion factor 14
Depth specification 3, 30
Detail of refractor definition 29–30
Determination of an optimum *XY* value 32–33, 67

Evjen equation, the 16
Examination of traveltime curves 31

Field data requirements 54
Field example 59–81
Field procedures for ambiguity 37

Generalized time-depth 13
  several *XY* values 32

Hales's method 14
Hidden layer (masked layer) 37
  maximum thickness 38

Intercept time 1, 52
  conventional 52
  generalized 52
Interpretation 56
  editing phase of 55
  recording with colors 56
Interpretation routine 55–57
Irregular ground surface 22, 25
Irregular refractor surface 18, 25, 74
Iterative method 2

Limits of definition 30
Linear velocity function 16

Masked layer (hidden layer) 37
Mean, migrated forward and reverse delay times 13
Method of differences 13
Methods for recognizing surface irregularities 31
Minimum equivalent refractor 30

Optimum *XY* spacing 2
Optimum *XY* value 2
  depth calculation 34–35
  unrecorded layers 34–35
Parabolic function results 16
Plus-minus method 7
  plus term 13
Processing routine 2

Raypath parameters 3
Reciprocal times 49–52
  corrections for errors 50
  distant shots 49
  errors 50–52
Redefinition of intercept time 52
Refractor surface, construction of 3
Refractor velocity 7, 74–81
  analysis 2
  plane layer case 7

Scott (1973) 2, 7
Second events 38
Selection of *XY* values 31
Specifying depths 3, 30
Strike of the refractor 3
Surface irregularities 31
Synthetic models 17–30, 90–91

Time-depth 13–16, 83
calculation of 2
definition of 13
near shotpoints 15
Time section 53–54
uniqueness of 53
Time series analysis 13–16, 83
Traveltime 37–39, 55
crosscorrelation methods 55
statistical methods 55
Traveltime curves, separation of distinctive features 33
Traveltime data, synthetic models 90–91
Traveltime expressions 3
multiple plane dipping layer case 5
Triple valued traveltime curves 38

Undetected layers 37–39, 83
recognized through *XY* values 39

Velocities corrected for dip 11
Velocity analysis functions 7
Velocity inversion 39
Velocity stratification 37

Wavelength considerations 57
Welcome Reef data 81, 93–101
Wiechert-Herglotz-Bateman integral 38